Multiple-choice mathematics
for GCSE

Charles Plumpton, M.A., Ph.D.

*Formerly Moderator in Mathematics, University of London
School Examinations Department; formerly Reader in Engineer-
ing Mathematics, Queen Mary College, University of London;
Chairman of the London Group Joint Subject Committee
(Mathematics)*

Eva Shipton, B.Sc.

*Chief Examiner in A–O level Mathematics, University
of London School Examinations Department;
formerly Deputy Head, Owen's School, Potters Bar*

M
MACMILLAN
EDUCATION

First published 1986
Reprinted 1987

Published by
MACMILLAN EDUCATION LTD
Houndmills, Basingstoke, Hampshire RG21 2XS
and London
Companies and representatives
throughout the world

Printed in Hong Kong

British Library Cataloguing in Publication Data
Plumpton, C.
Multiple choice mathematics for GCSE.
1. Mathematics—Examination, questions, etc.
I. Title   II. Shipton, E.
510'.76   QA43
ISBN 0—333—41695—3

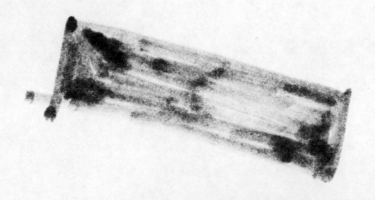

Contents

Preface

In the new GCSE Mathematics examination which is destined to take over from the GCE and CSE examinations in 1988, some Boards are including multiple-choice papers. The present book has the object of helping teachers who are preparing pupils for this examination, by providing a selection of multiple-choice papers geared to each different level of the examination. The lengths of the papers are designed to fit into school periods (or double periods in the case of the longer papers). Most teachers will be aware of the National Criteria for Mathematics which lay down minimum syllabuses for each level. The scheme of this book is as follows:

Tests 1 to 10 on List 1 of the National Criteria,
Tests 11 to 15 on List 2 of the National Criteria,
Tests 16 to 18 on List 2 plus additional material which the majority of Boards include for Level Y of the examination,
Tests 19 to 24 on Level Z, covering topics included by most Boards at this level.

It will be noticed that the book includes both 4-option and 5-option questions. All Level X questions are 4-option (except Tests 6 and 10, which are 5-option), at Level Z all are 5-option and the Level Y tests contain some of each.

Apart from the value of getting pupils accustomed to the multiple-choice type of question, teachers should find this book valuable in providing comprehensive revision during the final year of the course. One of the great merits of this type of question is the possibility of complete syllabus coverage in a single paper. Teachers will often find it useful to spend a lesson or two going over a paper in class. They will find that weaknesses of comprehension are revealed by the pupils' responses and time can then be spent on the weaker topics. For a more detailed study and appraisal of multiple-choice questions, teachers are recommended to see *Multiple-choice mathematics* by C. Plumpton (Macmillan Education, 1981). This book also provides papers suitable for Level Z pupils.

Eva Shipton
Charles Plumpton

Test 1

Time allowed: 40 minutes

1. A number divisible by 3 is

   A 413

   B 513

   C 613

   D 713

2. $1\frac{1}{8} - \frac{1}{4} =$

   A $1\frac{3}{8}$

   B $1\frac{1}{2}$

   C $1\frac{1}{8}$

   D $\frac{7}{8}$

3. $3 - 0\cdot14 - 0\cdot1 =$

   A $2\cdot96$

   B $2\cdot86$

   C $2\cdot85$

   D $2\cdot76$

4. $0\cdot07 \times 0\cdot08 =$

   A $0\cdot56$

   B $0\cdot056$

   C $0\cdot0056$

   D $0\cdot00056$

5. The cost of $5\frac{1}{2}$ kg of potatoes at 14p per kg is

   A 154p

   B 77p

   C 70p

   D 67p

6. The number of tiles, each a square of side 10 cm, needed to cover a rectangular area 3 m by 4 m is

   A 12

   B 120

   C 1200

   D 12 000

7. The ratio of 8p to £2, in its simplest terms, is

   A 4 : 1

   B 2 : 5

   C 8 : 200

   D 1 : 25

8. A scale of 4 cm to represent 5 km is equivalent to

   A 0·8 cm to 1 km

   B 1·25 cm to 1 km

   C 0·08 cm to 1 km

   D 80 mm to 1 km

9. Which *one* of the following is an obtuse angle?

   A 250°

   B 300°

   C 93°

   D 185°

1

10. Which of the following could *not* be the sizes of the angles of a triangle?

A 20°, 50°, 110°

B 40°, 40°, 100°

C 32°, 69°, 79°

D 54°, 16°, 120°

11. $12\frac{1}{2}\%$ of £8 =

A 25p

B £1

C £6.40

D £100

12. The number of millimetres in 1 km is

A 1000

B 10 000

C 100 000

D 1 000 000

13. The number 0·098 549 correct to 4 decimal places is

A 0·0985

B 0·098 55

C 0·0986

D 0·099

14. $S = a(4h + a)$.
When $a = 6$ and $h = 2$, $S =$

A 84

B 72

C 24

D 14

15.

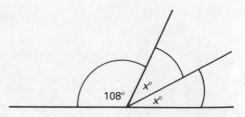

$x =$

A 36

B 41

C 72

D 82

16. For the numbers 1, 1, 2, 2, 3, 3, 3, 5, 5, 5, 5, 7 the difference between the median and the mode is

A 0·5

B 1·5

C 2·0

D 2·5

17. On a timetable using the 24 hour clock a time of 25 minutes past midnight appears as

A 12 25

B 23 35

C 24 25

D 00 25

18. A man stands facing East and then turns through 65° clockwise. He is now facing in the direction bearing

A 065°

B 155°

C 165°

D 295°

2

19. A girl makes a round Christmas cake of diameter 14 cm.

Taking $\pi = \dfrac{22}{7}$ the length, in cm, of the band she must buy just to encircle the cake is

A   22

B   44

C   88

D   154

20. A bag contains 3 red balls, 6 green balls and 2 white balls. A boy draws a ball at random from the bag. The probability that it will be white is

A $\dfrac{2}{11}$

B $\dfrac{2}{9}$

C $\dfrac{3}{11}$

D $\dfrac{1}{6}$

Test 2

Time allowed: 40 minutes

1. Which one of the following is *not* a square number?

    A 36

    B 49

    C 121

    D 160

2. 91 is a multiple of

    A 11

    B 7

    C 17

    D none of these, 91 is prime.

3. $2\frac{1}{3} + 3\frac{5}{6}$

    A $6\frac{1}{6}$

    B $5\frac{2}{3}$

    C $5\frac{1}{6}$

    D $3\frac{1}{6}$

4. 75p expressed as a fraction of £3 is

    A $\dfrac{1}{4}$

    B $\dfrac{1}{2}$

    C $\dfrac{3}{4}$

    D $\dfrac{4}{1}$

5. $(0\cdot1)^{3^r} =$

    A $0\cdot3$

    B $0\cdot01$

    C $0\cdot003$

    D $0\cdot001$

6. When the exchange rate is 12 francs to the £, for £7.50 I get, in francs,

    A 9

    B 84.6

    C 90

    D 96

7. A school starts at 9.05 a.m. To arrive in time, a boy whose journey takes 35 minutes must leave home, at the latest, by

    A 08 20

    B 08 30

    C 08 15

    D 08 25

8.

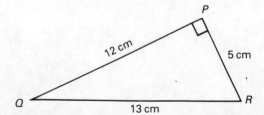

The area, in cm², of triangle *PQR* is

    A 30

    B $32\frac{1}{2}$

    C 60

    D 78

4

9. A photo measuring 3 cm by 4 cm is enlarged. The length of the enlargement is 10 cm. Its width is

A 7·5 cm

B 6·5 cm

C 6 cm

D 4·5 cm

10. A man has to pay tax of 30% on earnings of £4500. His tax bill is

A £135

B £150

C £1350

D £1500

11.

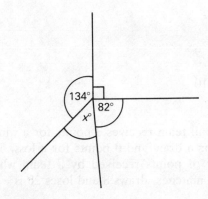

The point P could be

A (2, 0)

B (0, 2)

C (−2, 0)

D (0, −2)

12. The volume, in cm³, of a cuboid with adjacent edges of lengths 4 cm, 5 cm, 6 cm is

A 15

B 26

C 60

D 120

13.

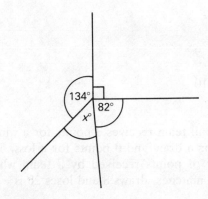

x =

A 64

B 56

C 54

D 46

14. When expressed correct to 2 significant figures, 9·2 × 0·3 =

A 2·7

B 2·8

C 2·76

D 28

15. The bearing of P from Q is 045°. The bearing of Q from P is

A 045°

B 135°

C 225°

D 325°

16.

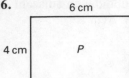

The ratio of the area of rectangle P to the area of rectangle Q is

A $5 : 6$

B $6 : 5$

C $10 : 9$

D $9 : 10$

17. A football team receives 3 points for a win, 1 point for a draw and 0 points for a loss. The number of points received by a team which wins 12 matches, draws 8 and loses 20 is

A 40

B 44

C 56

D 64

18.

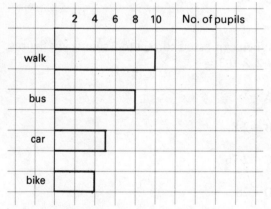

The bar chart shows the methods by which members of form 4A get to school. The probability that a member of 4A, picked at random, walks to school is

A $\dfrac{1}{3}$

B $\dfrac{10}{17}$

C $\dfrac{10}{27}$

D $\dfrac{4}{27}$

19. The sides of a rectangle, measured correct to the nearest centimetre, are 6 cm and 10 cm. The smallest possible value of the perimeter is

A 30 cm

B 30·04 cm

C 31 cm

D 32 cm

20. The formula for the length, L m, of a racing track is

$$L = 2\ell + \pi d.$$

When $\ell = 120$ m, $d = 50$ m and $\pi = 3 \cdot 14$, $L =$

A 157 m

B 240 m

C 277 m

D 397 m

Test 3

Time allowed: 40 minutes

1. $30^2 =$

    A 60

    B 90

    C 600

    D 900

2. The number of prime numbers between 30 and 40 is

    A 1

    B 2

    C 3

    D 4

3. On the 24 hour clock 20 00 hours is

    A 8 a.m.

    B 8 p.m.

    C 4 a.m.

    D 10 p.m.

4. $2 - \frac{7}{8} + \frac{1}{2} =$

    A $\frac{5}{8}$

    B $1\frac{3}{8}$

    C $1\frac{5}{8}$

    D $2\frac{3}{8}$

5. Which *one* of the following fractions is not equal to the others?

    A $\frac{15}{20}$

    B $\frac{10}{15}$

    C $\frac{12}{18}$

    D $\frac{30}{45}$

6. $7 - (2\cdot5 - 1\cdot8) =$

    A $2\cdot7$

    B $3\cdot7$

    C $6\cdot3$

    D $7\cdot3$

7. Which *one* of the following ratios is not equal to the ratio 2 : 5?

    A 6 : 15

    B 1 : 2·5

    C 3 : 7·5

    D 4 : 7

8. 15% VAT is added to a restaurant bill of £20. The total bill is

    A £23

    B £21.50

    C £35

    D £20.15

9. The scale of a map is 1 : 150 000. The actual distance represented by 3 cm on the map is

A 450 m

B 500 m

C 4·5 km

D 45 km

10. Which one of the following is the closest approximation to the area of a rectangle with sides of length 4·2 m and 5·8 m?

A 24 m^2

B 20 m^2

C 10 m^2

D 20 m

11.

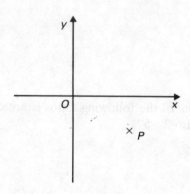

The coordinates of point P could be

A (3, 2)

B (−3, 2)

C (2, −3)

D (3, −2)

12.

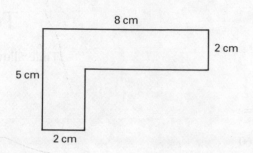

The area in cm^2, of the given figure is

A 40

B 22

C 18

D 16

13.

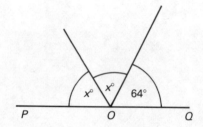

POQ is a straight line. $x =$

A 45

B 58

C 68

D 116

14.

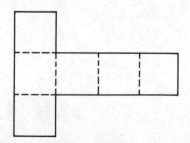

The diagram is the net of a

A cube

B cuboid

C triangular prism

D square based pyramid

8

15.

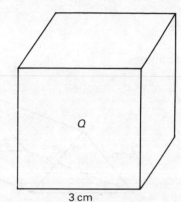

1 cm   3 cm

The ratio of the volume of cube P to the volume of cube Q is

A 1 : 3

B 1 : 6

C 1 : 9

D 1 : 27

16. A man started from a point P, walked 10 m due North, then turned and walked 10 m due West to a point Q. The bearing of Q from P is

A 045°

B 135°

C 225°

D 315°

17. Two angles of a triangle are 36° and 54°. The triangle is

A acute-angled

B isosceles

C right-angled

D obtuse-angled

18.

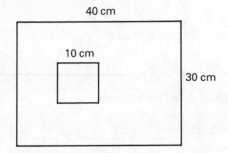

40 cm

10 cm

30 cm

The diagram shows a rectangular sheet of paper with a 10 cm square marked on it. In a competition a girl, blindfolded, sticks a pin in the paper and wins if her pin pierces inside the 10 cm square. The probability of her winning is

A $\dfrac{1}{12}$

B $\dfrac{1}{11}$

C $\dfrac{1}{3}$

D $\dfrac{1}{4}$

19. In the formula $v = u + at$, when $u = 35$, $a = 4\cdot2$ and $t = 20$, $v =$

A 43·4

B 59·2

C 119

D 129

20. A teacher cuts up a roll of material for a needlework lesson. She cuts 2 pieces of 2·5 m length, 4 pieces of 3 m length and 4 pieces of 3·5 m length. The average or mean length of a piece is

A $3\frac{1}{3}$ m

B 3·1 m

C 3 m

D 2·1 m

9

Test 4

Time allowed: 40 minutes

1. $1 \cdot 03 + 0 \cdot 087 + 31 \cdot 2 =$

   A $5 \cdot 02$

   B $1 \cdot 429$

   C $32 \cdot 217$

   D $32 \cdot 317$

2. 1011 is exactly divisible by

   A   17

   B   13

   C   11

   D   3

3. $1\frac{3}{10} - \frac{7}{15}$

   A $\frac{5}{6}$

   B $1\frac{4}{5}$

   C $1\frac{23}{30}$

   D $1\frac{5}{6}$

4. The square root of 30 is

   A   between 10 and 11

   B   between 5 and 6

   C   between 6 and 7

   D   between 14 and 16

5. $\frac{1}{2}$ of $3\frac{1}{4} =$

   A $1\frac{1}{8}$

   B $1\frac{3}{8}$

   C $1\frac{5}{8}$

   D $6\frac{1}{2}$

6.

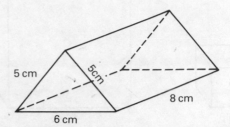

The hollow prism without ends, shown in the diagram, is constructed from a rectangular piece of card whose dimensions, not allowing for joins, are

   A   8 cm by 6 cm

   B   8 cm by 10 cm

   C   8 cm by 11 cm

   D   8 cm by 16 cm

7. Of the price of a dress, 33% is the retailer's profit, 38% the manufacturer's profit, 25% the cost of manufacture, leaving the remainder of the price for the cost of materials which is

   A   4%

   B   14%

   C   86%

   D   96%

8.

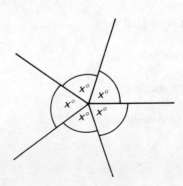

$x =$

   A   72

   B   60

   C   45

   D   36

9. In the formula $t = \dfrac{v - u}{a}$ when $v = 45$, $u = 28$, $a = 10$,

$t =$

A 0·17

B 1·7

C 2·7

D 42·8

10. When $5 = 11 - 2x$, $x =$

A 8

B 6

C 3

D −3

11.

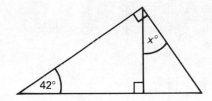

$x =$

A 132

B 58

C 48

D 42

12. The number of squares, each of side 1 cm, which can be fitted into a square of side 1 m is

A 100

B 1000

C 10 000

D 100 000

13.

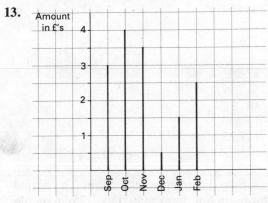

The chart shows a girl's savings per month for 6 months of the year. The mean savings per month are

A £2

B £2.50

C £3

D £3.50

14. $\dfrac{60\cdot4 \times 3\cdot9}{120\cdot2}$ is approximately

A 2

B 0·2

C 20

D 1·5

15.

The difference in hours shown on the two clocks when both are a.m. times on the same day, is

A $5\frac{1}{2}$

B $6\frac{1}{2}$

C $7\frac{1}{2}$

D 17

11

16. 8% of £90 =

A 72p

B £7.20

C £8

D £11.25

17. On a map 1 cm represents 5 km. The actual distance, in km, represented by 1·25 cm on the map is

A 0·2

B 0·25

C 6·25

D 7·25

18. £1 = $1.15. Then $2300 =

A £2000

B £200

C £2465

D £246.50

19. In a football pool 3 points are given for a score draw, 2 points for a no-score draw, 1½ points for an away win and 1 point for a home win. Of 8 matches chosen, 4 are score draws, 1 is a no-score draw, 1 is an away win and the rest are home wins. The number of points scored is

A $16\frac{1}{2}$

B $17\frac{1}{2}$

C 18

D $18\frac{1}{2}$

Questions **20** to **22** relate to the graph showing the relationship between temperatures given in the Fahrenheit scale (*F*) and those given in the Celsius scale (*C*).

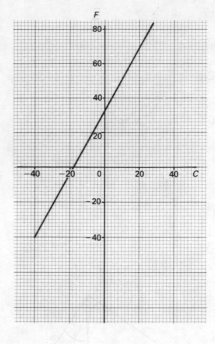

20. When *F* = 14, *C* =

A −10

B −8

C 10

D 58

21. When *C* = 20, *F* =

A 64

B 68

C −7

D 7

22. *C* = *F* when *C* =

A 40

B 32

C −18

D −40

12

Questions **23** to **25** relate to the bar chart showing the grades achieved by pupils in an examination.

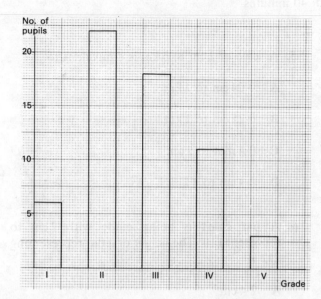

23. The number of pupils who sat the exam was

A 20

B 50

C 60

D 100

24. The modal grade achieved was

A I

B II

C III

D IV

25. The percentage of pupils achieving grade I was

A 12%

B 11⅔%

C 10%

D 6%

Test 5

Time allowed: 40 minutes

1. 540 m =

  A 54 km

  B 5·4 km

  C 0·54 km

  D 0·054 km

2. Two angles of a triangle are 30° and 60°. The triangle is

  A equilateral

  B obtuse-angled

  C isosceles

  D right-angled

3. 0·005 =

  A $\dfrac{1}{50}$

  B $\dfrac{1}{20}$

  C $\dfrac{1}{200}$

  D $\dfrac{1}{2000}$

4. A man uses a pie diagram to show how his weekly expenditure of £72 is allocated. The angle of the sector representing expenditure of £30 on food is

  A 300°

  B 150°

  C 75°

  D 30°

5. A scale of 2 cm to 5 km is equivalent to

  A 0·04 cm to 1 km

  B 0·25 cm to 1 km

  C 0·4 mm to 1 km

  D 0·4 cm to 1 km

6. A bar chart is being made with 2 cm to represent £10. The height of a bar representing £12 is

  A 2·4 cm

  B 1·2 cm

  C 2·2 cm

  D 6 cm

7. The square root of 1000 is a number

  A between 10 and 20

  B between 30 and 40

  C between 100 and 110

  D between 300 and 400

8. 18 is a multiple of

  A 2, 3, 4 and 6

  B 2, 3, 6 and 9

  C 2, 3, 6 and 12

  D 3, 6, 8 and 9

9. $\frac{1}{2}$ of $1\frac{1}{3}$ =

A $\frac{1}{3}$

B $\frac{1}{2}$

C $\frac{2}{3}$

D $\frac{5}{6}$

10. Tracy earns £60 a week, from which her employer deducts £3.20 for National Insurance and £6.15 for PAYE. Tracy's 'take-home' pay is

A £50.65

B £50.75

C £51.65

D £57.05

11. A table is made to convert °Celsius to °Fahrenheit.

Celsius	30	20	10	0	−10	−20
Fahrenheit	86	68	50	32	?	?

The two missing values are

A 22, 12

B 14, 4

C −14, −4

D 14, −4

12. Three angles of a quadrilateral are 72°, 49°, 88°. The size, in degrees, of the remaining angle is

A 141

B 151

C 161

D 61

13.

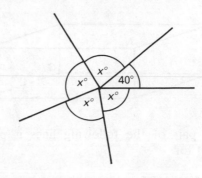

x =

A 80

B 72

C 60

D 40

14. The most accurate geometrical description of this page of the book is that it is a

A quadrilateral

B parallelogram

C rectangle

D square

15.

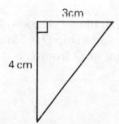

The diagram shows the base of a right prism of volume 150 cm³. The height of the prism is found by dividing its volume by the area of its base. The height, in cm, of the prism is

A 12·5

B 15

C 20

D 25

16.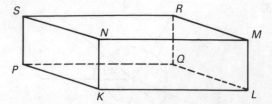

Each pair of the following lines is parallel except for

A *PK* and *QL*

B *PS* and *LM*

C *SR* and *KL*

D *PQ* and *RN*

17. A bicycle wheel of diameter 50 cm makes exactly 100 complete turns in travelling, without skidding, along a line drawn on the ground. Assuming that $\pi = 3 \cdot 14$, the length, in m, of the line is

A 15·7

B 78·5

C 157

D 785

18. A coastguard station *P* is 5 km due West of a lighthouse *L*, and a ship *Q* is 5 km due South of *P*. The bearing of *Q* from *L* is

A 315°

B 225°

C 135°

D 45°

19. An interest of £24 per annum on £250 is a rate of

A 6%

B 9·6%

C 12%

D 24%

20. A gardener bought a packet of mixed sweet pea seeds. The packet contained 25 seeds and the mixture consisted of 6 white-, 4 red-, 5 blue- and the rest violet-flowered plants. The probability that a seed chosen at random will give a violet-flowered plant is

A $\dfrac{2}{5}$

B $\dfrac{3}{5}$

C $\dfrac{2}{3}$

D $\dfrac{1}{5}$

21.

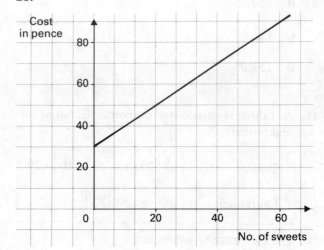

The graph shows the relation between the cost of a box of sweets and the number of sweets the box contains. The number of sweets a 50p box contains is

A 20

B 40

C 50

D 80

16

22.

Golders Green	11 46
Mill Hill	12 02
Shenley	12 20

Above is an extract from a coach timetable. The time to get from Golders Green to Shenley is

A 16 minutes

B 18 minutes

C 34 minutes

D 56 minutes

23.

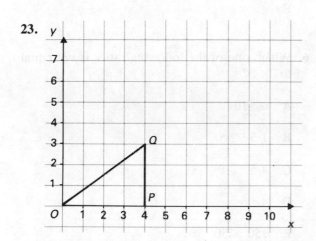

The triangle OPQ is transformed into the triangle $OP'Q'$ by an enlargement with centre O and scale factor 2. The coordinates of Q' are

A (8, 0)

B (6, 8)

C (8, 6)

D (0, 8)

24. Each day a school meals centre uses 12 kg of potatoes per 100 pupils. 30 kg of potatoes would be just sufficient for

A 150 pupils

B 180 pupils

C 240 pupils

D 250 pupils

25. A box of internal measurements 10 cm by 20 cm by 30 cm contains children's bricks, each of which is a cube of side 2 cm. The number of bricks the box can hold is

A 750

B 1500

C 3000

D 6000

17

Test 6

1. A number divisible by 7 is

   A 81

   B 91

   C 101

   D 111

   E 121

2. An estimate, made by rounding off the numbers in the calculation

   $$\frac{0 \cdot 51 \times 101}{0 \cdot 19}$$

   is

   A 2·5

   B 25

   C 50

   D 250

   E 500

3. $0 \cdot 2 \times 0 \cdot 03 =$

   A 6

   B 0·6

   C 0·06

   D 0·006

   E 0·0006

4. $\frac{3}{8} =$

   A 80%

   B $37\frac{1}{2}\%$

   C 30%

   D $26\frac{2}{3}\%$

   E $12\frac{1}{2}\%$

5. $1\frac{1}{3} - \frac{1}{2} =$

   A $\frac{2}{3}$

   B $\frac{3}{2}$

   C $\frac{1}{6}$

   D 6

   E $\frac{5}{6}$

6. Which one of the following ratios is *not* equal to 3 : 5?

   A 6 : 10

   B 1·5 : 2·5

   C 4·5 : 7·5

   D 4 : 6

   E 150 : 250

7. The number of tiles, each a square of side 5 cm, needed to cover exactly a rectangular area 2 m by 3 m is

   A 2400

   B 1200

   C 240

   D 120

   E 24

8. A rectangle has length x cm and breadth y cm. The perimeter, in cm, is

   A $(2x + 2y)$

   B $(x + y)$

   C xy

   D $2xy$

   E $4xy$

18

9.

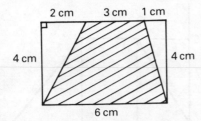

The area, in cm², of the shaded region is

A 12

B 16

C 18

D 20

E 24

10. A girl is working values of y for different values of x in the formula

$$y = \frac{x - 3}{4}.$$

She makes a table but she gets one value of y wrong. Which?

x	3	3·2	5·1	5·4	6
y	0	0·5	0·525	0·6	0·75

A 0

B 0·5

C 0·525

D 0·6

E 0·75

11. When a man drives in town, his car does 24 miles per gallon of petrol. The number of miles of town driving which he can do on a full tank of 12 gallons is

A 2

B 240

C 248

D 278

E 288

12. A photo measuring 2 cm by 4 cm is enlarged. Given that the length of the enlargement is 10 cm, then the width is

A 4 cm

B 4·5 cm

C 5 cm

D 6 cm

E 8 cm

13.

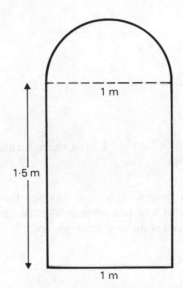

An honours board consists of a rectangle surmounted by a semi-circle as shown. Taking $\pi = 3·14$, the length, in m, of edging needed to go round the outside of the board is

A 5·57

B 5·64

C 6·57

D 7·14

E 8·14

14. 6% of £2400 is

A £400

B £144

C £14.40

D £4

E £1.44

15.

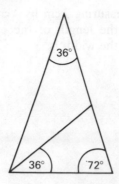

The number of *isosceles* triangles in the given diagram

A   is 0

B   is 1

C   is 2

D   is 3

E   cannot be found from the information given.

16. A man puts £500 into a Savings Bank at an interest of 8% per annum. At the end of the first year his money amounts to

A   £900

B   £580

C   £540

D   £508

E   £40

17. The difference between the mean and the mode of the numbers 1, 2, 3, 3, 5, 6, 7, 8, 13 is

A $\frac{1}{3}$

B $1\frac{1}{3}$

C   2

D $2\frac{1}{3}$

E $2\frac{1}{2}$

18.

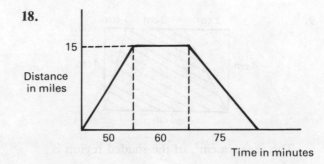

The travel graph shown is for a cyclist's journey from town *A* to town *B*, 15 miles away, and back again after a stop for lunch. His speeds, in miles/h, on the outward and return journeys respectively were

A   18, 12

B   18, 20

C $12\frac{1}{2}$, $18\frac{3}{4}$

D $12\frac{1}{2}$, 12

E   20, $18\frac{3}{4}$

19. A month is to be chosen at random from an ordinary calendar. The probability that it will have 30 days is

A $\frac{7}{12}$

B $\frac{4}{7}$

C $\frac{4}{11}$

D $\frac{1}{2}$

E $\frac{1}{3}$

20. Given that $S = 3 \cdot 1\, r^2$, the value of S when $r = 3$ is

A   9·3

B   18·6

C   27·9

D   81·9

E   86·49

20

21.

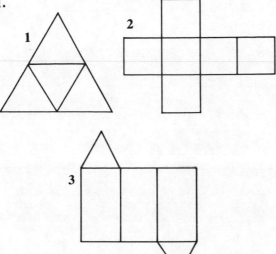

Which of the above could be the net of a solid figure?

A 1 only

B 2 only

C 1 and 2 only

D 2 and 3 only

E 1, 2 and 3

22.

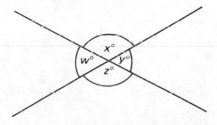

Which of the following is (are) true?
(1) $x + y = 180$ (2) $x = z$
(3) $x + y + z + w = 360$

A (1) only

B (2) only

C (3) only

D all of them

E none of them

The diagrams apply to questions **23** to **25**.

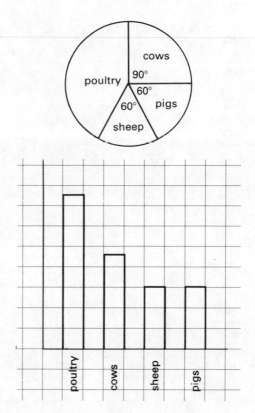

The two diagrams show the number of animals kept by a farmer. There are 180 cows, and the height of the 'sheep' column in the second diagram is 3 cm.

23. The number of poultry is

A 120

B 150

C 180

D 240

E 300

24. The height of the 'cows' column in the second diagram is

A 2 cm

B 4·5 cm

C 5 cm

D 6 cm

E 7·5 cm

25. The total number of animals kept is

A   720

B   660

C   600

D   420

E   360

Test 7

Time allowed: 1 hour

1. The only one of the following numbers which is not prime is

   A 31

   B 71

   C 91

   D 101

2. A number divisible by 3 is

   A 112

   B 233

   C 466

   D 729

3. The LCM of 6, 9 and 18 is

   A 3

   B 18

   C 36

   D 54

4. A train leaves a station at 3.20 p.m. On the timetable, which uses the 24 hour clock, this time would be shown as

   A 03 20

   B 13 20

   C 14 20

   D 15 20

5.

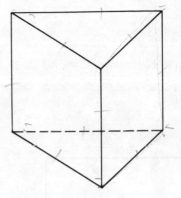

The diagram shows a regular solid which has the following number of edges

   A 5

   B 6

   C 9

   D 18

6.

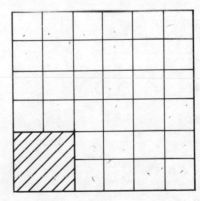

The fraction of the diagram shaded is

   A $\frac{1}{3}$

   B $\frac{1}{2}$

   C $\frac{1}{6}$

   D $\frac{1}{9}$

7. A child's bricks are cubes of side 5 cm. The number of bricks which can be fitted into a rectangular box whose inside measurements are 20 cm by 15 cm by 10 cm is

A 12

B 24

C 120

D 300

8.

The area, in cm², of the given figure is

A 30

B 34

C 42

D 44

9. $0.3 \times 0.015 =$

A 0.45

B 0.045

C 0.0045

D 0.0015

10. $3^4 - 2^3 =$

A 4

B 5

C 6

D 73

11. A regular solid can be constructed by folding along the thickened lines in all but one of the following nets. Which will *not* form a solid?

A

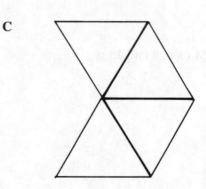

B

C

D

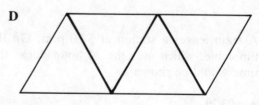

12. The mean of 3, 5, 7, 7, 8 is

A 7

B 6

C 5

D 4

24

13. The smallest of the following fractions is

A $\dfrac{4}{15}$

B $\dfrac{13}{60}$

C $\dfrac{1}{5}$

D $\dfrac{1}{4}$

14. $\frac{1}{8} =$

A 0·25

B 0·15

C 0·0125

D 0·125

15.

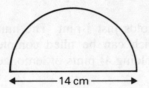

Taking $\pi = \dfrac{22}{7}$, the length, in cm, of the perimeter of the given diagram is

A 36

B 44

C 58

D 154

16.

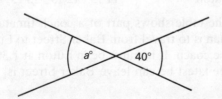

$a =$

A 40

B 50

C 60

D 140

17.

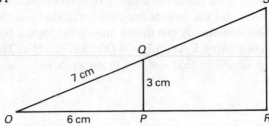

Triangle *ORS* is the enlargement of triangle *OPQ* with scale factor 2. *PR* =

A 6 cm

B 7 cm

C 12 cm

D 14 cm

18.

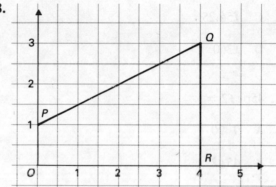

The area, in square units, of *OPQR* is

A 4

B 8

C 12

D 16

19. In a word game the letters are marked on small tiles and the vowels are kept separate from the consonants. A girl draws one vowel from a bag containing 5 A's, 5 E's, 4 O's, 3 E's, 3 U's. The probability that she will draw an A is

A $\dfrac{1}{5}$

B $\dfrac{1}{2}$

C $\dfrac{1}{3}$

D $\dfrac{1}{4}$

20. The interest on £450 invested for 1 year at 8% per annum is

A £36

B £44

C £56.25

D £360

21. 45p : £2 =

A 9 : 20

B 45 : 2

C 9 : 50

D 9 : 40

22. 4p expressed as a percentage of £1 is

A 0·04%

B 0·4%

C 4%

D 40%

23. A sale offers goods at a 15% discount. The discount on a dress of marked price £40 is

A 60p

B £6

C £15

D £46

24. A boy has £x and spends y pence. The amount, in pence, left is

A $x - y$

B $100(x - y)$

C $100x - y$

D $x - \dfrac{y}{100}$

25. 4 litres is approximately equal to 7 pints. The nearest equivalent, in pints, to 28 litres is

A 16

B 35

C 49

D 56

26. A cup holds just $\frac{1}{3}$ pint. The number of such cups which can be filled completely from a bottle holding $4\frac{1}{2}$ pints of lemonade is

A 12

B 13

C 14

D 15

27.

Victoria	09 22	10 37	11 19	12 37
Baker Street	09 33	10 48	11 30	12 48
London Colney	10 28	11 28	12 28	13 28
St. Albans	10 36	11 38	12 36	13 36
Luton	11 15	12 15	13 15	14 15

The table shows part of a coach timetable. A man is to travel from Baker Street to Luton by the coach. If he must be in Luton at 1.30 p.m., the latest he can leave Baker Street is

A 11 19

B 11 30

C 10 48

D 09 33

28.

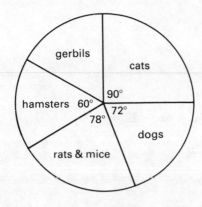

The diagram shows the numbers of pets kept by a class of primary school children. Given that there were 12 dogs, the number of gerbils was

A 10

B 11

C 12

D 60

29. One angle of an isosceles triangle is 100°. Another angle of the triangle is

A 40°

B 50°

C 80°

D 100°

30. At midnight one night the temperature recorded in London was −4°C. At noon the following day the temperature recorded was 7°C. The rise in temperature from midnight to noon was

A 11 °C

B 7 °C

C 4 °C

D 3 °C

Test 8

Time allowed: 1 hour

1. The *only one* of the following which is a prime number is

   A 27

   B 49

   C 51

   D 61

2. Which *one* of the following is a square number?

   A 8

   B 24

   C 40

   D 49

3. $1\frac{1}{6} - \frac{2}{3} =$

   A 1

   B $1\frac{1}{2}$

   C $\frac{1}{2}$

   D $\frac{1}{3}$

4. $\frac{5}{6}$ of £3 =

   A £2.50

   B £2

   C £1.50

   D 50p

5. $x + 3x + 6x - 2x =$

   A 12x

   B 8x

   C 7x

   D 8

6. The area, in cm², of a square of side 1 m, is

   A 100

   B 1000

   C 10 000

   D 100 000

7.

The reading shown on the meter dials is

   A 1070

   B 1089

   C 1179

   D 1079

8. $2^5 - 5^2 =$

   A 0

   B 7

   C 22

   D 57

28

9. When *N* indicates North, the bearing 130° is indicated by *OP* in

A

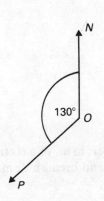

B

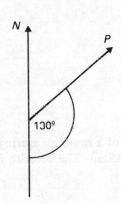

C

D

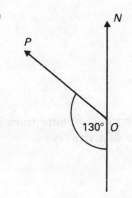

10. $2 - 0.325 =$

A 1·785

B 1·675

C 1·775

D 0·675

11. $2\frac{1}{2}$ kg at 28p per kg =

A 56p

B 60p

C 70p

D 80p

12. $(0·3)^2 =$

A 0·6

B 0·9

C 0·09

D 0·009

13. A number exactly divisible by 7 is

A 107

B 119

C 121

D 131

14. 15p as a fraction of £3 =

A $\dfrac{1}{20}$

B $\dfrac{1}{15}$

C $\dfrac{3}{20}$

D $\dfrac{20}{1}$

15.

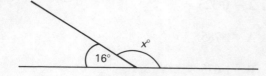

$x =$

A 174°

B 164°

C 84°

D 74°

16. Two angles of a triangle are 36° and 42°. The triangle is

A equilateral

B isosceles

C acute-angled

D obtuse-angled

17. 8 mm : 2 m =

A 2 : 5

B 4 : 1

C 1 : 25

D 1 : 250

18.

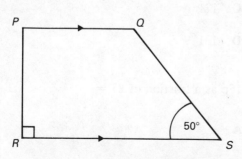

Angle $Q =$

A 130°

B 90°

C 50°

D 40°

19. 12% of £75 =

A 16p

B £16

C 90p

D £9

20. 15% VAT is added to a bill of £60. The amount actually paid is

A £9

B £51

C £69

D £75

21. The perimeter, in m, of a rectangle, which is of length $3x$ m and breadth $2x$ m, is

A $10x$

B $5x$

C $6x^2$

D $12x^2$

22. In a plan of a room measuring 6 m by 4 m, the length is 15 cm. The breadth is

A 1 cm

B 2·5 cm

C 10 cm

D 12·5 cm

23. A man starts on a 45 minute train journey at 10 58. His time of arrival is

A 11 03

B 11 33

C 11 43

D 12 03

24.

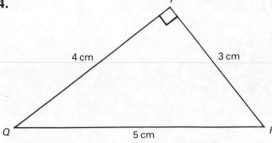

The area, in cm², of triangle PQR is

A 6

B 12

C 10

D 7½

25. A girl draws a single card from an ordinary pack of 52 playing cards. The probability of her drawing an ace is

A $\dfrac{1}{52}$

B $\dfrac{1}{26}$

C $\dfrac{1}{13}$

D $\dfrac{1}{4}$

The diagram below applies to both questions **26** and **27**. It shows amounts collected for a form's charity during a special week's effort.

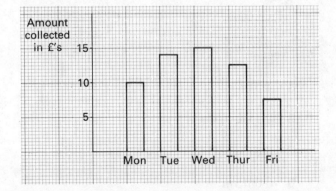

26. An amount of £12.50 was collected on

A Monday

B Tuesday

C Thursday

D Friday

27. The total amount collected during the week was

A £15

B £49

C £58

D £59

The diagram below applies to both questions **28** and **29**. It shows the distance–time graph for a cyclist going from town P to town Q and then returning later.

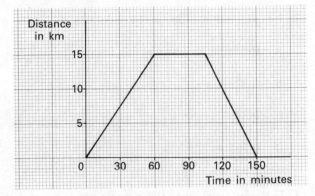

28. The time spent in town Q was

A 45 minutes

B 1 h

C 1 h 15 minutes

D 1 h 30 minutes

29. The cyclist's speed returning from Q to P was

A 20 km/h

B 15 km/h

C 11¼ km/h

D 10 km/h

30. The circumference, in cm correct to the nearest whole number, of a circle of radius 7 cm is

A 22

B 44

C 154

D 308

Test 9

1. 76p + £1.42 + £2 =

   A £2.20

   B £3.18

   C £4.18

   D £11.02

2. $\frac{3}{8}$ expressed as a decimal is equal to

   A 0·4

   B 0·375

   C 0·37

   D 0·365

3. Estimate, correct to the nearest whole number
$3·98 \times 2·07$.

   A 6

   B 8

   C 9

   D 12

4. The prime factors of 18 are

   A $2 \times 3 \times 3$

   B 2×9

   C 3×6

   D 1×18

5. $1·705 - 0·63 - 0·1 =$

   A 0·975

   B 1·075

   C 1·175

   D 1·641

6. $3\frac{5}{6} + 1\frac{1}{4} =$

   A $2\frac{7}{12}$

   B $4\frac{1}{12}$

   C $4\frac{3}{5}$

   D $5\frac{1}{12}$

7. $0·07 \times 0·03 =$

   A 0·0021

   B 0·021

   C 0·21

   D 2·1

8. $1\frac{3}{10} - \frac{7}{15} =$

   A $1\frac{5}{6}$

   B $1\frac{23}{30}$

   C $1\frac{4}{5}$

   D $\frac{5}{6}$

9. 15 cm =

   A 1·5 m

   B 0·15 m

   C 0·015 m

   D 0·0015 m

10. The highest common factor of 48 and 60 is

   A 6

   B 12

   C 18

   D 240

11. 24p expressed as a fraction of £3 is

   A $\dfrac{8}{1}$

   B $\dfrac{6}{25}$

   C $\dfrac{2}{25}$

   D $\dfrac{2}{250}$

12. $\frac{2}{3}$ of $1\frac{1}{8}$ =

   A $\frac{1}{12}$

   B $\frac{3}{8}$

   C $\frac{3}{4}$

   D $\frac{4}{3}$

13. 25 lb of potatoes cost £1.25. At the same rate, 3 lb of potatoes cost

   A 5p

   B 15p

   C 25p

   D 30p

14. The angles of a triangle are $x°$, $2x°$, $3x°$.
$x =$

   A 30

   B 60

   C 90

   D 45

15.

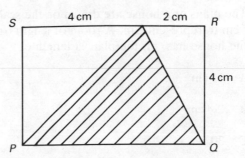

PQRS is a rectangle. The area, in cm², of the shaded region is

   A 24

   B 16

   C 12

   D 8

16. A point moves in a straight line so that its distance, x metres, from a fixed point O at time t seconds is given by the formula

$$x = 2 + 15t.$$

At time 3 seconds the distance of the point from O is

   A 2 m

   B 17 m

   C 45 m

   D 47 m

17. A wheel is turning about its centre, which is fixed, so that it completes 1 revolution in 6 seconds. In 1 second it turns through an angle of

   A 15°

   B 30°

   C 45°

   D 60°

18. The plans of a house are drawn on the scale of 5 cm to represent 2 m. A room of length 6 m in the house has, on the plan, a length of

A 15 cm

B 2·4 cm

C 30 cm

D 8 cm

19.

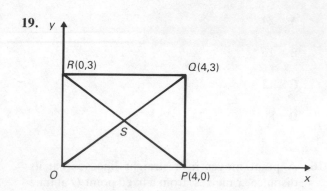

The coordinates of S are

A (2,2)

B (2,3)

C (2,1½)

D (0,2)

20.

Pocket money	50p	75p	£1	£1.25	£1.50
No. of children	4	5	10	7	4

A class of 30 children make the above table showing the pocket money they receive (in each case rounded to the nearest 25p). They make a column graph in which the height of the 50p column is 6 cm. The height of the £1 column is

A 15 cm

B 12 cm

C 10·5 cm

D 7·5 cm

21. The median of 5, 2, 7, 5, 8, 4, 7, 7, 6, 11, 5 is

A 4

B 5

C 6

D 7

22. A man walks 15 miles at 5 miles per hour and then 3 miles at 3 miles per hour. His average speed, in miles per hour, for the whole journey is

A 4½

B 4

C 3¾

D 3½

23. One angle of a quadrilateral is equal to 120° and the other three angles are equal. Each is equal to

A 60°

B 80°

C 90°

D 120°

24.

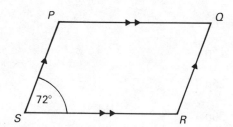

$PQRS$ is a parallelogram with $PQ\|SR$ and $PS\|QR$.
Angle $Q =$

A 18°

B 36°

C 72°

D 108°

25. 18% of £2.50 =

   A 7·2p

   B 18p

   C 36p

   D 45p

26. 10% of a form of 30 children were absent one day. How many children were present?

   A 3

   B 10

   C 20

   D 27

27. An investment of £400 earns £15 interest in 6 months. The rate per cent per annum is

   A 15%

   B $7\frac{1}{2}$%

   C 30%

   D $3\frac{3}{4}$%

28.

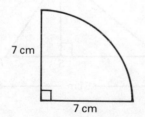

7 cm

7 cm

In the figure the region shown is a quadrant of a circle. Taking $\pi = \dfrac{22}{7}$, its perimeter, in cm, is

   A 22

   B 25

   C 36

   D 44

29. A jar weighs x g when empty and it holds 30 g of spice. The weight, in g, of 12 full jars is

   A $360x$

   B $12x + 30$

   C $30x + 12$

   D $12x + 360$

30. A dress priced at £50 was reduced by a 10% discount in a sale and then by a further 10% of the new price when it did not sell. Its final price was

   A £39.50

   B £40

   C £40.50

   D £45

Test 10

Time allowed: 1 hour

1. The only number of the following which is *not* a prime is

   A 7

   B 17

   C 27

   D 37

   E 47

2. $1 - 0.125 =$

   A 0·985

   B 0·975

   C 0·885

   D 0·875

   E 0·125

3. The number of edges of a triangular prism is

   A 3

   B 4

   C 6

   D 8

   E 9

4. When 70 is divided in the ratio $3:7$ the two parts are

   A 30, 70

   B 30, 40

   C 21, 49

   D 21, 59

   E $17\frac{1}{2}$, $52\frac{1}{2}$

5.

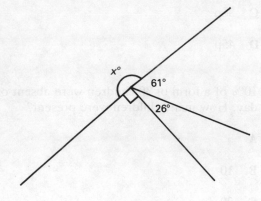

   $x =$

   A 273

   B 193

   C 183

   D 103

   E 93

6.

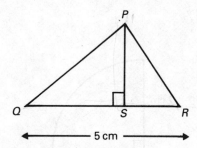

   Area $\triangle PQR = 6\,\text{cm}^2$.
   $PS =$

   A 1·2 cm

   B 1·5 cm

   C 2·4 cm

   D 3 cm

   E 4 cm

7. 25p as a fraction of £1 is

A $\frac{1}{8}$

B $\frac{1}{4}$

C $\frac{1}{2}$

D $\frac{4}{1}$

E $\frac{1}{25}$

8. The largest of the following fractions is

A $\frac{5}{12}$

B $\frac{3}{8}$

C $\frac{17}{48}$

D $\frac{1}{3}$

E $\frac{9}{24}$

9. $7\frac{1}{2}\%$ of £50 is

A 6p

B 15p

C £3.75

D £37.50

E none of these

10. $S = a(4h + a)$.
When $a = 3$, $h = 2$, $S =$

A 11

B 28

C 33

D 42

E 60

11.

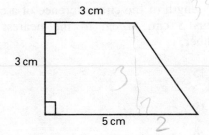

The area of the given figure, in cm², is

A 9

B 12

C 15

D 24

E 45

12. The best approximation to $2\cdot7 \times 3\cdot01$ is

A 0·6

B 0·8

C 8

D 80

E 800

13. Two angles of a triangle are of sizes 19° and 25°. The size of the third angle is

A 44°

B 46°

C 56°

D 136°

E 146°

14. The length of the circumference of a circle of radius 5 cm, in cm to the nearest whole number, is

A 15

B 16

C 30

D 31

E 78

15. On a 24 hour clock the time 12.15 p.m. appears as

A 12 15

B 23 15

C 23 45

D 24 15

E 00 15

16. The median of the numbers 2, 6, 8, 4, 3, 8, 8, 4, 5, 8, 9 is

A 3

B 4

C 5

D 6

E 8

17. When the exchange rate is 12 francs to the £, for £10.50 I get, in francs,

A 12.60

B 120.60

C 126

D 125

E 120

18.

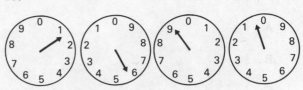

The meter reading shown is

A 1590

B 1500

C 1501

D 1591

E 2691

19.

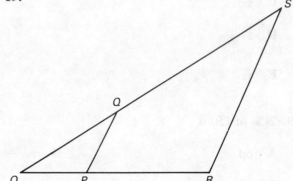

Triangle *ORS* is the enlargement of triangle *OPQ* with scale factor 3.

$$\frac{OP}{PR} =$$

A $\frac{2}{1}$

B $\frac{1}{2}$

C $\frac{1}{3}$

D $\frac{3}{1}$

E $\frac{1}{4}$

20.

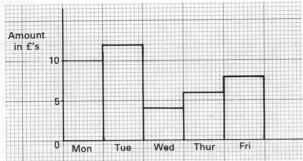

The bar chart shows the takings of a school's tuck shop. The total week's takings amounted to

A £12

B £13

C £15

D £30

E £40

21. A man stands facing West and then turns through 45° clockwise. He is now facing in a direction whose bearing is

A 045°

B 135°

C 145°

D 225°

E 315°

22. A bag contains 3 red, 2 white, 4 green and 6 yellow balls. When a ball is drawn at random from the bag, the probability that it will not be a yellow ball is

A $\dfrac{2}{5}$

B $\dfrac{3}{5}$

C $\dfrac{4}{5}$

D $\dfrac{2}{3}$

E $\dfrac{1}{3}$

23.

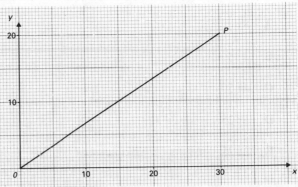

The coordinates of the middle point of OP are

A (15, 10)

B (30, 20)

C (10, 6·5)

D (10, 15)

E (20, 13)

24. A model engine of length 12 cm is built to a scale of 1 : 50. The actual length of the engine is

A 6 m

B 60 m

C 600 m

D 2·4 m

E 0·24 m

25. A man starts on a 45 minute journey at 10 56. His time of arrival is

A 12 01

B 11 41

C 11 31

D 11 10

E 11 01

39

26. The ratio of 16p to £2 in its simplest terms is

A   8 : 1

B   4 : 5

C   4 : 50

D   1 : 25

E   2 : 25

27. The mean of the numbers 2, 3, 3, 3, 4, 4, 5, 5, 7, 9 is

A   3

B   3·5

C   4

D   4·5

E   5

28. The total cost, in pence, of x pencils at 8p each and y ballpoints at 10p each is

A $80xy$

B $18xy$

C $18(x+y)$

D $8x + 10y$

E $10x + 8y$

29. The number of cm^3 in $1\,m^3$ is

A   100

B   1000

C   10 000

D   100 000

E   1 000 000

30. Given that 8 pints = 4·5 litres, the cost of 1 pint of wine at £4 per litre is

A   50p

B   £2.25

C   £2.50

D   £8

E   £18

Test 11

1. $\frac{1}{4} \div \frac{1}{3} =$

   A $\frac{3}{4}$

   B $\frac{4}{3}$

   C $\frac{1}{12}$

   D $\frac{7}{12}$

2. $0.2 \div 7$, correct to 3 significant figures, is

   A 0.029

   B 0.0286

   C 0.0285

   D 0.286

3.

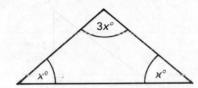

   $x =$

   A 18

   B 30

   C 36

   D 108

4. Each exterior angle of a regular 9-sided polygon is equal to

   A 30°

   B 36°

   C 40°

   D 140°

5. The smallest of 4 consecutive even numbers is x. The largest number is

   A $x + 3$

   B $x + 4$

   C $x + 5$

   D $x + 6$

6. $(3x + 2y) - (x - y) =$

   A $2x + 3y$

   B $2x + y$

   C $4x + 3y$

   D $3 + 2y$

7. The area, in cm², of a triangle of base $2x$ cm and height x cm is

   A $2x^2$

   B x^2

   C $3x/2$

   D x

8. The area, in cm², of a semi-circle of radius 6 cm is

   A 9π

   B 12π

   C 18π

   D 36π

41

9. A man's weekly wage increases by 10% per year. Which one of the following graphs could best represent this fact?

A
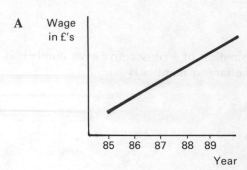

B
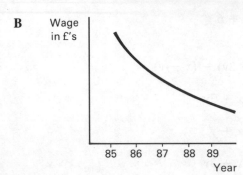

C
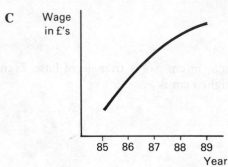

D
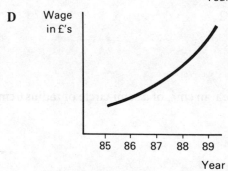

10. The mean weight of 4 women is 60 kg. The mean weight of 3 of these women is 64 kg. The weight, in kg, of the other woman is

A   48

B   58

C   62

D   66

11.

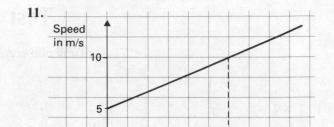

The diagram shows the time–speed graph of a car. During the interval $0 \leqslant t \leqslant 12$, the distance, in m, moved by the car is

A   30

B   60

C   90

D   120

12.

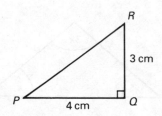

sin P =

A $\dfrac{3}{4}$

B $\dfrac{3}{5}$

C $\dfrac{4}{5}$

D $\dfrac{4}{3}$

13. The gradient of the line $x + 2y = 3$ is

A -2

B $-\frac{1}{2}$

C $\frac{1}{2}$

D $\frac{3}{2}$

14. A rectangle enlarged by scale factor 2 has its sides increased by

A 200%

B 150%

C 100%

D 50%

15. The tax on dividends is 30%. Given that the tax paid on a gross dividend of £G is £66, then G =

A £19.80

B £85.80

C £198

D £220

16.

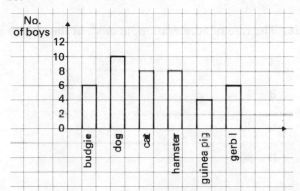

A survey was held in a junior boys' school of the pets kept by the children. It was found that all the boys who took part in the survey had at least one pet and some of them had two pets. None had more than two pets. Given that 30 boys took part in the survey, the number keeping two pets was

A 42

B 12

C 8

D 6

17. An approximation of $\dfrac{\sqrt{(50)} \times 15 \cdot 9}{28 \cdot 16}$, correct to 1 significant figure, is

A 0·4

B 4

C 14

D 29

18. $1 \cdot 76 \times 10^{-3} =$

A 0·00176

B 0·0176

C 0·176

D 1760

19.

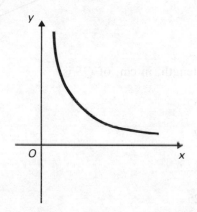

The diagram could be part of the graph of

A $y = x + 1$

B $y = x^2$

C $y = -x$

D $y = \dfrac{1}{x}$

20. Which *one* of the following is a rational number?

A $\sqrt{30}$

B $\sqrt{20}$

C $\sqrt{18}$

D $\sqrt{9}$

43

21. A man walks so that he keeps two landmarks, P and Q, in view and is always equal distances from the two. He walks along

  A a circle with centre P or Q

  B a straight line through P and Q

  C a straight line bisecting PQ at right angles

  D a straight line parallel to PQ

The figure relates to questions **22, 23** and **24**.

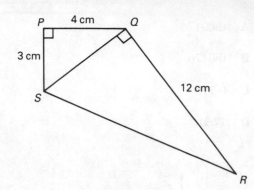

22. The length, in cm, of QS is

  A 5

  B 7

  C 12·5

  D 25

23. The length, in cm, of SR is

  A 19

  B 17

  C $\sqrt{193}$

  D 13

24. The area, in cm², of the quadrilateral $PQRS$ is

  A 6

  B 30

  C 36

  D 72

The figure relates to questions **25, 26, 27** and **28**. It shows the time–speed graph of a train between stations P and Q.

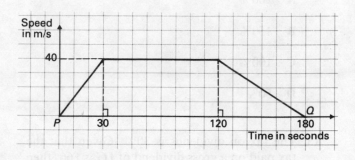

25. The speed, in m/s, of the train 15 s after leaving P is

  A 40

  B 30

  C 20

  D 10

26. The rate of acceleration, in m/s², of the train is

  A $\frac{2}{3}$

  B $\frac{4}{3}$

  C 40

  D 80

27. The time, in s, taken for the train to go from P to Q is

  A 60

  B 90

  C 120

  D 180

28. The distance, in m, from P to Q is

  A 7200

  B 5400

  C 3600

  D 540

44

Questions **29** and **30** relate to a solid cylinder of radius 5 cm and height 8 cm.

29. The volume, in cm^3, of the cylinder is

  A 200π

  B 100π

  C 40π

  D 25π

30. The total surface area, in cm^2, of the cylinder is

  A 130π

  B 105π

  C 90π

  D 80π

Test 12

1. 24 300, when expressed in standard form, is

   A $2 \cdot 43 \times 10^3$

   B $2 \cdot 43 \times 10^4$

   C $0 \cdot 243 \times 10^5$

   D $24 \cdot 3 \times 10^3$

2. Each interior angle of a regular octagon is equal to

   A 144°

   B 135°

   C 108°

   D 45°

3. The HCF and LCM of 16 and 40 are respectively

   A 4 and 80

   B 8 and 10

   C 8 and 640

   D 8 and 80

4. Expressed as a decimal fraction, $\frac{5}{8}$ =

   A 0·5

   B 0·6

   C 0·625

   D 1·6

5. An approximate value of $\dfrac{0 \cdot 608}{\sqrt{(35 \cdot 8)} \times 1 \cdot 9}$, working to 1 significant figure, is

   A 0·05

   B 0·06

   C 0·5

   D 0·6

6. Goods bought for £20 are sold for £24. The percentage profit on the cost price is

   A $16\frac{2}{3}\%$

   B 20%

   C $83\frac{1}{3}\%$

   D 120%

7. Three children aged 8, 10 and 12 receive £15 to be divided among them in the ratios of their ages. The youngest gets

   A £4

   B £5

   C £6

   D £8

8. Given that $x = \dfrac{pq}{r}$, then q =

   A $\dfrac{px}{r}$

   B $\dfrac{p}{rx}$

   C $\dfrac{x}{pr}$

   D $\dfrac{rx}{p}$

9.

x	-1	0	1	2
y	9	0	3	12

In the table of values given for the function $y = 3x^2$, one of the y values is wrong. Which?

A 9

B 0

C 3

D 12

10. $5xy^3 - 5xy =$

A $5xy(y^2)$

B $5xy(xy^2 - 1)$

C y^2

D $5xy(y^2 - 1)$

11. When $7 - 3x = 10 + x$,
$x =$

A $-1\frac{1}{2}$

B $-1\frac{1}{3}$

C $-\frac{3}{4}$

D $\frac{3}{4}$

12. The following costs of manufacturing a car are represented on a pie chart. Materials: £2400; Labour: £1600; Overheads: £800. The angle representing Labour is

A 30°

B 60°

C 120°

D 180°

13.

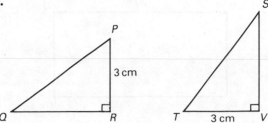

In which one of the following cases could the pair of triangles given *not* be congruent?

A $PQ = ST$

B $\angle Q = \angle S$

C $PQ = SV$

D $\angle P = \angle T$

14.

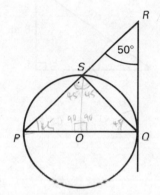

In the diagram POQ is a diameter and QR is a tangent. Angle $PQS =$

A 50°

B 45°

C 40°

D 30°

15. A man starts from a point O and walks 10 km due West and then 9 km due South to a point P. The bearing of P from O is approximately

A 042°

B 132°

C 138°

D 228°

16.

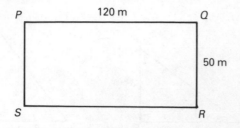

PQRS is a rectangular field. The shortest distance, in m, from *P* to *R* is

A $\sqrt{(11\,900)}$

B $\sqrt{(6000)}$

C   130

D   1300

17.

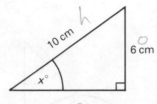

$\cos x° =$

A   0·6

B   0·75

C   0·8

D   1·25

18. A darts player scores 21, 44, 57 and 62 in successive 'visits' to the dartboard. What must he score in his fifth 'visit' if the mean score for the five visits is 53?

A   46

B   53

C   60

D   81

19. In a circle of radius 10 cm the angle of a sector is 45°. The area, in cm², of the sector is

A   2·5π

B   12·5π

C   25π

D   100π

20. A coin is to be tossed twice. The probability that the first toss gives a head and the second a tail is

A $\dfrac{3}{4}$

B $\dfrac{1}{2}$

B $\dfrac{1}{3}$

D $\dfrac{1}{4}$

21. A bicycle wheel is of diameter 50 cm. Taking π as 3·14, the number of *complete* turns made by the wheel in travelling 100 m without skidding is

A   636

B   63`

C   31

D   6

Questions **22–24** refer to the speed–time graph of a tube train between 2 stations.

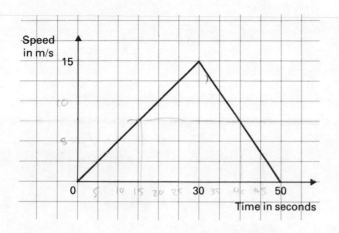

22. The train was accelerating for

A 30 s

B 20 s

C 15 s

D 50 s

23. The total distance, in m, travelled was

A 225

B 375

C 450

D 750

24. The speed of the train was greater than $7\frac{1}{2}$ m/s for

A $7\frac{1}{2}$ s

B 10 s

C 15 s

D 25 s

Questions **25–27** refer to the table showing the amount at Compound Interest on £100 for various rates per cent per annum.

Year \ Rate	5%	6%	7%	8%
1	£105	£106	£107	£108
2	£110.25	£112.36	£114.49	£116.64
3	£115.76	£119.10	£122.50	£125.97
4	£121.55	£126.25	£131.08	£136.05

25. £100 at 6% for 3 years amounts to

A £106

B £115.76

C £119.10

D £11910

26. The interest rate for £200 to amount to £245 in 3 years is

A 6%

B 7%

C $7\frac{1}{2}$%

D 15%

27. The sum which will amount to £655.40 after 4 years at 7% is

A £5000

B £600

C £500

D £50

49

Questions **28–30** refer to the diagram of a square tile patterned with a circle and an inner square. Take π as 3·14 in these questions.

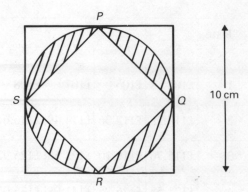

28. The area, in cm^2, of the circle in the pattern is

A 78·5

B 100

C 157

D 314

29. The length, in cm, of *PQ* is

A $\sqrt{10}$

B $\sqrt{50}$

C 10

D $\sqrt{200}$

30. The total area, in cm^2, of the shaded parts is

A 28·5

B 25

C 21·5

D 14·25

Test 13

1. The highest common factor of 36 and 48 is

   A 3

   B 6

   C 9

   D 12

2. The value of 1·0949 correct to 2 decimal places is

   A 1·09

   B 1·10

   C 1·094

   D 1·095

3. The lengths of the sides of a rectangle measured to the nearest cm are 3 cm and 5 cm. The smallest possible value of the perimeter is

   A 14 cm

   B 14·04 cm

   C 15 cm

   D 15·5 cm

4. $\dfrac{0·6 \times 0·2}{0·01} =$

   A 0·12

   B 1·2

   C 12

   D 120

5. Which *one* of the following is *not* equal to any of the others?

   A $\frac{3}{10}$

   B 30 : 110

   C 30p : £1

   D 30%

6. A shop uses 1200 units of electricity in 15 days. 2000 units of electricity should be sufficient for just

   A 9 days

   B 12 days

   C 25 days

   D 80 days

7. When the exchange rate is £1 = 11·5 francs and £1 = 2300 lire, the number of lire obtained for 1 franc is

   A 2

   B 20

   C 200

   D 2000

8. Including VAT a man's restaurant bill came to £40. Adding service charge at $12\frac{1}{2}\%$ to this, the total paid was

   A £52.50

   B £45

   C £40.25

   D £35

9. When $x = -3$ and $y = -2$,
$x^2 - 2y =$

A   13

B   −13

C   −10

D   −5

10. When $2x - 3 = 6$, the value of x is

A   1·5

B   4·5

C   6

D   9

11. Which of the following is (are) true?
1 The x-axis is the line $y = 0$.
2 The y-axis is the line $x = 0$.
3 The line $y = 2x - 1$ passes through the point (1,1).

A   **1** and **2** only

B   **3** only

C   **1** only

D   **1**, **2** and **3**

12. A ship is on course 320° and turns 90° to the right. Its new course is

A   040°

B   050°

C   230°

D   220°

13. The size, in degrees, of the exterior angle of a regular 9-sided polygon is

A   20

B   40

C   140

D   160

14.

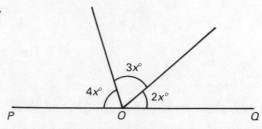

POQ is a straight line. $x =$

A   10

B   20

C   30

D   40

15. One angle of an isosceles triangle is 98°. Another angle of the triangle

A   is 98°

B   is 82°

C   is 41°

D   cannot be found; there is insufficient information to find another angle

16. In a circle of area 60 cm^2 the angle of a sector is 45°. The area, in cm^2, of the sector is

A   7·5

B   10

C   15

D   30

17. A hall, 60 m by 24 m, is shown in a scale plan to have length 15 cm. The width, shown in the plan is

A   6 cm

B   12 cm

C   24 cm

D   96 cm

18. A coin is to be tossed twice. The probability that the first toss gives a head and the second a tail is

A $\frac{3}{4}$

B $\frac{1}{2}$

C $\frac{1}{3}$

D $\frac{1}{4}$

19.

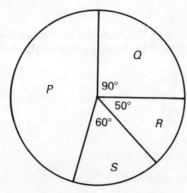

The pie diagram shows the division of a total of £3600. Sector *P* represents

A £1600

B £900

C £600

D £160

20. The mean mass of five men is 70 kg. One of them is of mass 86 kg. The mean mass, in kg, of the remaining four is

A 74

B 70

C 66

D 46

21. £270 is divided into 3 amounts in the ratios 2 : 3 : 4. The largest amount is

A £30

B £54

C £60

D £120

22.

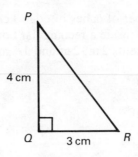

sin *R* =

A $\frac{4}{5}$

B $\frac{3}{5}$

C $\frac{4}{3}$

D $\frac{3}{4}$

23. The vertices of a triangle are the points $(-1,0)$, $(5,0)$ and $(3,5)$. Its area, in square units, is

A 9

B 10

C 15

D 30

24.

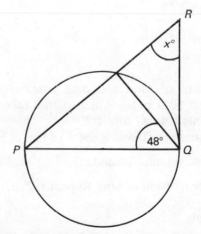

PQ is a diameter of the circle and *QR* is a tangent.
x =

A 42

B 48

C 52

D 58

53

25. The number of cubes of edge 1 cm which may be packed inside a rectangular box, with inside measurements 2 m, 2 m and 1·5 m, is

A $\quad 6 \times 10^9$

B $\quad 6 \times 10^5$

C $\quad 6 \times 10^6$

D $\quad 6 \times 10^3$

26. Given that $x = 8$ and $y = \frac{1}{2}$, which one of the following has the greatest value?

A $\quad x + y$

B $\quad x^y$

C $\quad xy$

D $\quad \dfrac{x}{y}$

27. In triangle PQR, angle $Q = 36°$, angle $R = 90°$, $PQ = 5$ cm.
The length, in cm, of QR is

A $\quad 5 \cos 36°$

B $\quad 5 \sin 36°$

C $\quad 5 \tan 36°$

D $\quad 5 \cos 54°$

28. A mixture of sand and peat contains 40% sand and 60% peat. Which of the following statements is (are) correct?
1 The quantity of sand is $66\frac{2}{3}$% of that of peat.

2 $\frac{2}{5}$ of the mixture is sand.

3 The proportion of sand to peat is $2:3$.

A $\quad$ **1** only

B $\quad$ **1** and **2** only

C $\quad$ **2** and **3** only

D $\quad$ **1, 2** and **3**

29. The diameter of a cylindrical lawn roller is 30 cm and its width is 40 cm. The area, in cm², of the lawn which the roller covers in one complete revolution is

A $\quad 1200$

B $\quad 1200\pi$

C $\quad 2400\pi$

D $\quad 9000\pi$

30. A house which cost £32 000 was later sold at a loss of 5%. The selling price of the house was

A $\quad$ £30 400

B $\quad$ £31 995

C $\quad$ £33 600

D $\quad$ none of these

Test 14

Time allowed: 45 minutes

1. The lowest common multiple of 6, 8 and 12 is

   A   24

   B   36

   C   48

   D   576

2. $0.04 \div 0.5 =$

   A   8

   B   0.8

   C   0.08

   D   0.008

3. $(-2) \times (+3) \times (-4) =$

   A   3

   B   −9

   C   −24

   D   +24

4. The number 0.095 47, correct to 3 significant figures, is

   A   0.095

   B   0.0955

   C   0.0954

   D   0.0965

5. $3\frac{1}{3} \div \frac{1}{3} =$

   A $1\frac{1}{9}$

   B   3

   C   9

   D   10

6. The price paid for a £60 coat which had been reduced by 15% in a sale was

   A   £69

   B   £51

   C   £45

   D   £40

7. When £72 is divided in the ratio 1 : 3, the two parts are

   A   £6 and £66

   B   £12 and £60

   C   £18 and £54

   D   £24 and £48

8. A car covers 108 km in 2 hours 15 minutes. The average speed, in km/h, of the car is

   A   72

   B   60

   C   54

   D   48

9.

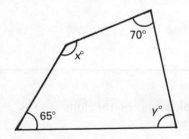

   $x + y =$

   A   45

   B   135

   C   180

   D   225

55

10. Both points (0, 2) and (0, 4) lie on the line

   A $x = 0$

   B $y = 0$

   C $y = 6$

   D $y + x = 6$

11.

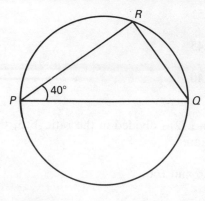

PQ is a diameter of the circle whose radius is 5 cm. $PR =$

   A 10 sin 40° cm

   B 5 cos 40° cm

   C 10 tan 50° cm

   D 10 cos 40° cm

12. The inside measurements of a box are 10 cm by 12 cm by 4 cm. The number of cubes of edge 2 cm which the box will hold is

   A 480

   B 240

   C 60

   D 30

13. Of which *one* of the following is $x = 1$ the solution?

   A $5x = 3x + 2$

   B $3x = x - 2$

   C $x/2 = 2$

   D $1 - x = 2$

14.

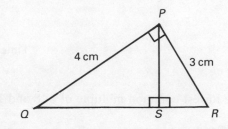

$PS =$

   A 1·2 cm

   B 1·8 cm

   C 2·4 cm

   D 3·6 cm

15.

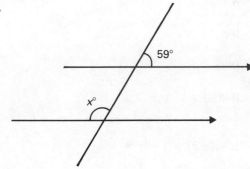

$x =$

   A 131

   B 121

   C 31

   D 59

16. When $x = 3$ and $y = -2$, $2x - y =$

   A 10

   B 8

   C 4

   D −8

17. A boy measured a length as 4 cm when it should have been 5 cm. His percentage error was

  A 20%

  B 25%

  C 80%

  D 125%

18. The exterior angle of a regular polygon could *not* be

  A 50°

  B 40°

  C 30°

  D 20°

19. Which one of the following is closest to the square root of 0·008?

  A 0·003

  B 0·03

  C 0·09

  D 0·009

20. The mean and median of the numbers 2, 5, 5, 5, 6, 6, 10, 17 are, respectively

  A 5, 6

  B 7, 5·5

  C 7, 5

  D 9·5, 5·5

21. The area, in cm², of a circle of diameter 6 cm is

  A 6π

  B 9π

  C 12π

  D 36π

22.

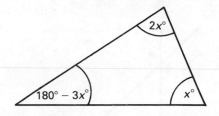

  A $x = 60$

  B $x = 30$

  C $x = 45$

  D x cannot be found without further information

23. The scale of a map is 1 : 150 000. A distance of 7·5 km on the ground is represented on the map by

  A 500 cm

  B 5 cm

  C 20 cm

  D 50 cm

24. When $\dfrac{x}{3} + 1 = 2$, $x =$

  A 9

  B 6

  C 3

  D $\frac{1}{3}$

25. A second-hand car was bought for £2000 and, each year, depreciates by 40% of its value at the beginning of that year. Its value at the end of two years is

  A £480

  B £720

  C £800

  D £1600

26.

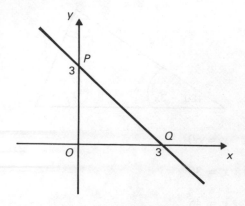

The equation of the line PQ is

A $\quad x + y = 3$

B $\quad x - y = 3$

C $\quad x = y$

D $\quad y = x + 3$

27. In a district where rates are levied at £1.20 in the £, the rate payable on a house of rateable value £300 is

A $\quad$ £36

B $\quad$ £250

C $\quad$ £360

D $\quad$ £3600

The diagram applies to questions **28, 29** and **30**. Tracy's income after tax amounts to £54 per week and she allocates it as shown in the pie chart.

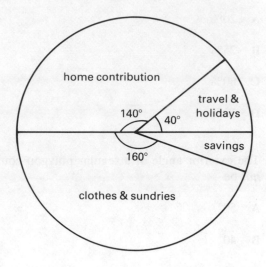

28. The fraction of her income which Tracy saves is

A $\quad \dfrac{1}{18}$

B $\quad \dfrac{1}{12}$

C $\quad \dfrac{1}{9}$

D $\quad \dfrac{17}{18}$

29. The weekly amount which Tracy contributes at home is

A $\quad$ £42

B $\quad$ £24

C $\quad$ £21

D $\quad$ £10.50

30. The ratio
amount spent on clothes and sundries : home contribution
is

A $\quad 4 : 9$

B $\quad 9 : 4$

C $\quad 7 : 8$

D $\quad 8 : 7$

Test 15

Time allowed: 45 minutes

1. 0·0046 when expressed in standard form is

   A $4·6 \times 10^{-2}$

   B $4·6 \times 10^{-3}$

   C $4·6 \times 10^{-4}$

   D 46×10^{-4}

2. $(0·06)^3 =$

   A 0·000 018

   B 0·000 216

   C 0·002 16

   D 0·0216

3. 38 ÷ 3, correct to 3 significant figures, is

   A 12·2

   B 12·6

   C 12·667

   D 12·7

4. A man left $\frac{2}{3}$ of his fortune to his wife and $\frac{1}{3}$ of the remainder to his son. Given that the man left a total of £63 000, the son got

   A £7000

   B £10 500

   C £14 000

   D £21 000

5. Which of the following numbers is (are) irrational?
   1 $\sqrt{2}$
   2 π
   3 3·142

   A All of them

   B 1 and 2 only

   C 1 only

   D 2 only

6. Which one of the following numbers cannot be a probability value?

   A 0

   B 0·5

   C 1

   D 5

7. A ship leaves port and steams on a bearing of 235° for a distance of 12 km. It is now further South of the port by a distance, in km, of

   A 12 sin 55°

   B 12 cos 35°

   C 12 cos 55°

   D 12 tan 35°

8. The interior angle of a regular 15-sided polygon is

   A 24°

   B 48°

   C 156°

   D 168°

9.

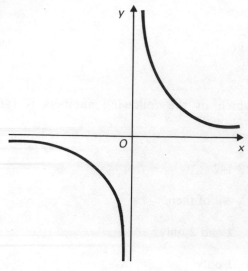

The curve shown could have equation

A $\quad y = \dfrac{1}{x}$

B $\quad y = \dfrac{1}{x^2}$

C $\quad y = 1 + \dfrac{1}{x}$

D $\quad y = x^2$

10.

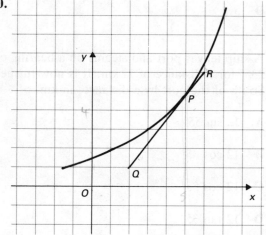

RQ is the tangent to the given curve at P. The gradient of the curve at P is

A $\quad \dfrac{4}{3}$

B $\quad \dfrac{4}{5}$

C $\quad \dfrac{3}{4}$

D $\quad \dfrac{5}{4}$

11.

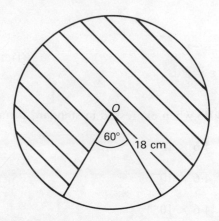

O is the centre of the given circle. The area, in cm^2, of the shaded region is

A $\quad 30\pi$

B $\quad 54\pi$

C $\quad 216\pi$

D $\quad 270\pi$

12.

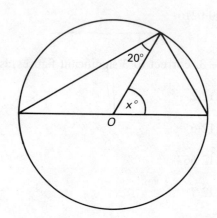

O is the centre of the given circle. $x =$

A $\quad 20$

B $\quad 35$

C $\quad 40$

D $\quad 70$

13. In a sale an item costing £2 is sold for £2.15. The percentage profit on cost price is

A $\quad 7\frac{1}{2}\%$

B $\quad 13\frac{1}{3}\%$

C $\quad 15\%$

D $\quad 30\%$

14. The formula $V = \pi r^2 h$ for the volume of a cylinder is transformed to find r in terms of V and h. The transformed formula is $r =$

A $\quad \sqrt{\left(\dfrac{V}{\pi h}\right)}$

B $\quad \sqrt{(V - \pi h)}$

C $\quad \dfrac{V}{2\pi h}$

D $\quad \frac{1}{2}(V - \pi h)$

15. When $x = 2$, $y = -3$, $z = 1$, $x + y - z =$

A $\quad 4$

B $\quad 2$

C $\quad 0$

D $\quad -2$

16. Two angles of a triangle are 56° and 68°. The triangle is

A $\quad$ equilateral

B $\quad$ isosceles

C $\quad$ obtuse-angled

D $\quad$ right-angled

17.

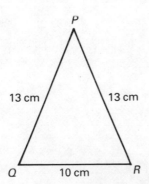

The area, in cm², of triangle PQR is

A $\quad 60$

B $\quad 65$

C $\quad 120$

D $\quad 130$

18. Points which are equidistant from the coordinate axes Ox and Oy lie on

A $\quad$ the line $y = x$ only

B $\quad$ the line $x = 0$ or the line $y = 0$

C $\quad$ the line $y = -x$ only

D $\quad$ the line $y = x$ or the line $y = -x$

19. The numbers of large, medium and small eggs sold by a retailer are in the ratios $4 : 5 : 3$. On a pie chart representing this data, the size of the angle of the sector representing the number of medium eggs sold would be

A $\quad 30°$

B $\quad 75°$

C $\quad 120°$

D $\quad 150°$

20. The diagonals of a kite are of lengths 6 cm and 8 cm. The length of the longest side of the kite

A $\quad$ is 10 cm

B $\quad$ is 6 cm

C $\quad$ is 5 cm

D $\quad$ cannot be found from the information given

21. When $15 - 2x = 3x$, $x =$

A $\quad 15$

B $\quad 6$

C $\quad 3$

D $\quad -3$

22.

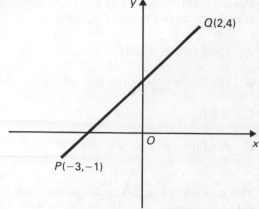

The middle point of the line PQ is

A $(\frac{1}{2}, 1\frac{1}{2})$

B $(-\frac{1}{2}, -1\frac{1}{2})$

C $(-\frac{1}{2}, 1\frac{1}{2})$

D $(-2\frac{1}{2}, -2\frac{1}{2})$

23. A triangle enlarged by scale factor 2 has its sides increased by

A 50%

B 100%

C 150%

D 200%

24.

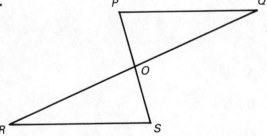

In the diagram $PO = OS$ and $QO = OR$.
Which of the following is (are) true?
1 $PQ = RS$
2 PQ is parallel to RS
3 triangle POQ = triangle ROS in area

A all of them

B **1** and **2** only

C **1** only

D **2** only

25. A train travelled for 20 minutes at 45 km/h and then for 40 minutes at 60 km/h. Its average (mean) speed, in km/h, for the journey was

A 15

B $52\frac{1}{2}$

C 55

D 105

26. The number of cylindrical cups of radius 3 cm and height 8 cm which can be filled from a cylindrical drum of liquid of radius 0·3 m and height 0·8 m is

A 10 000

B 1000

C 100

D 10

The given timetable for buses running between Barnet and St. Albans applies to questions **27** and **28**.

Barnet	07 45	08 00	08 14	08 31
St. Albans	08 30	08 45	08 59	09 16

27. A girl who lives in Barnet attends a school which is ten minutes walk from St. Albans bus station. The time of the last bus she can catch at Barnet to get her to school by twenty past nine is

A   07 45

B   08 00

C   08 14

D   08 31

28. Given that the distance on the bus route from Barnet to St. Albans is 10 miles, then the average scheduled speed, in miles/h, for the bus is

A $7\frac{1}{2}$

B   12

C $13\frac{1}{3}$

D   45

The given diagram applies to questions **29** and **30**.

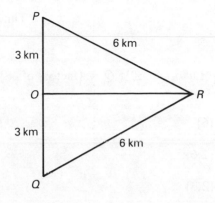

P, Q, R are points with P due North of Q and $PQ = QR = RP = 6$ km. O is the mid-point of PQ.

29. The distance, in km, of OR is

A   27

B $\sqrt{45}$

C $\sqrt{27}$

D   3

30. The bearing of Q from R is

A   120°

B   150°

C   210°

D   240°

63

Test 16

Time allowed: 45 minutes

1. P = {factors of 30}, Q = {factors of 48}.
$P \cap Q$ =

A {6}

B {1,6}

C {2,3}

D {1,2,3}

E {1,2,3,6}

2. The value of 32×0.4, correct to two significant figures, is

A 1·3

B 1·28

C 13

D 12·8

E 130

3. Given that $3x - 1 = 17$, then $6x + 1$ =

A 19

B 33

C 34

D 36

E 37

4. When 30% profit is made on a cost price of £2 the selling price is

A £2.60

B £2.30

C £2.15

D £1.40

E 60p

5. ½% is the same as

1 0·05

2 $\frac{1}{200}$

3 5×10^{-3}

Which of the following is (are) correct?

A 1 only

B 2 only

C 1 and 2 only

D 2 and 3 only

E 1, 2 and 3

6. P, Q, R are 3 sets in which $Q \subset R$, $P \cap Q = \varnothing$, $P \cap R \neq \varnothing$. The Venn diagram which best illustrates these facts is

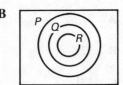

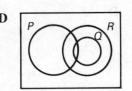

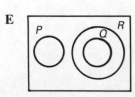

64

7. A regular polygon has each exterior angle equal to 45°. The number of sides of the polygon is

A   5

B   6

C   8

D   9

E   10

8.

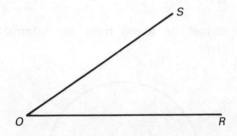

A man walks so that he keeps the same distance from the two lines OR and OS. The line he walks along is

A   the bisector of angle ROS

B   a bisector of line OR

C   a bisector of line OS

D   a line parallel to OR

E   a line parallel to OS

9. The total rateable value of a borough is £5 600 000 and the total expenditure for the year amounts to £8 400 000. The rate in the £ required to cover this is

A   67p

B   £1.50

C   £1.80

D   £3

E   £6.67

10. A kitchen floor is rectangular and measures 3 m by 5 m. The least number of square tiles, of edge 25 cm, needed to cover the floor is

A   60

B   75

C   240

D   375

E   400

11. As a result of a pay rise, employees who had previously been paid £W per week were given an increase of 7% plus £3. Their new weekly wage was, in £,

A $\dfrac{7W}{100} + 3$

B $\dfrac{107W}{100} + 3$

C $\dfrac{107(W + 3)}{100}$

D $\dfrac{100}{93}(W + 3)$

E $\dfrac{100W}{93} + 3$

12. Given that x and y satisfy equations
$$2x + y = -1,$$
$$x - y = 13,$$
then the value of x is

A -14

B -9

C $-4\frac{2}{3}$

D 4

E 12

13. Given that the point $(\frac{1}{3}, 3)$ lies on the curve $y = px^3$, then the value of p is

A $\dfrac{1}{81}$

B $\dfrac{1}{9}$

C 3

D 9

E 81

14. A cylindrical vessel of radius 5 cm holds 150 cm^3 of water when completely full. The height, in cm, of the vessel is

A $\dfrac{6}{\pi}$

B $\dfrac{\pi}{6}$

C $\dfrac{15}{\pi}$

D $\dfrac{\pi}{15}$

E $\dfrac{30}{\pi}$

15.

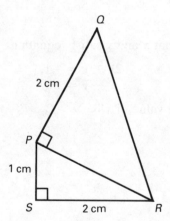

$QR =$

A $\sqrt{5}$ cm

B 3 cm

C 5 cm

D $\sqrt{29}$ cm

E 9 cm

16. In a tombola prizes are awarded for drawing any number ending in 0 or 5. If the total number of tickets in the drum is an exact multiple of 10, and all consecutive whole numbers are used, the probability of getting a prize

A is $\dfrac{1}{10}$

B is $\dfrac{1}{9}$

C is $\dfrac{1}{4}$

D is $\dfrac{1}{5}$

E cannot be found from the information given

17.

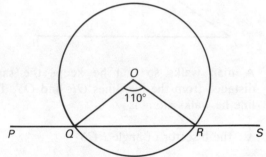

O is the centre of the circle. Angle $PQO =$

A 35°

B 70°

C 125°

D 135°

E 145°

18. Which of the following is (are) rational?
1 $\sqrt{121}$
2 22/7
3 $\sqrt{8}$

A All of them

B **1** and **2** only

C **2** only

D **2** and **3** only

E none of them

66

19.

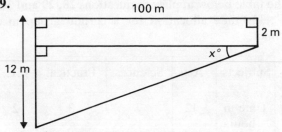

Which one of the following is true?

A $\sin x = 0 \cdot 1$

B $\tan x = 0 \cdot 1$

C $\sin x = 0 \cdot 12$

D $\tan x = 0 \cdot 12$

E $\sin x = 0 \cdot 02$

20. The lengths of the sides of two similar polygons are in the ratio $9 : 4$. The ratio of their areas is

A $3 : 2$

B $9 : 4$

C $81 : 16$

D $16 : 81$

E $729 : 64$

21.

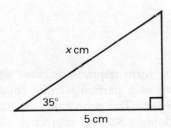

$x =$

A $\dfrac{5}{\sin 35°}$

B $5 \cos 35°$

C $5 \sin 35°$

D $\dfrac{5}{\cos 35°}$

E $5 \tan 35°$

22. $12\frac{1}{2}\%$ of 20% of £10 000 is

A   £3250

B   £2500

C   £1250

D   £250

E   £25

23.   2, 3, 3, 4, 4, 5, 5, 5, 6, 7
The median value of the given numbers is

A   5

B $4\frac{2}{3}$

C $4 \cdot 5$

D $4 \cdot 4$

E   3

24. The exchange rates for various currencies on a certain day were given as: £1 = 11.76 francs, £1 = 2470 lire, £1 = 167 drachmas, £1 = 1.265 dollars. Which of the following is (are) true?

1 123.5 lire = 5p
2 £1 > 1 dollar
3 1 franc < 10 drachmas

A   **1** and **2** only

B   **2** and **3** only

C   **1** only

D   **2** only

E   all of them

The figure below applies to questions **25, 26** and **27**.

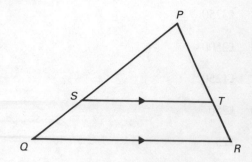

$PS : SQ = 2 : 1$ and ST is parallel to QR.

25. $ST : QR =$

   A $1 : 2$

   B $2 : 3$

   C $2 : 1$

   D $3 : 2$

   E $3 : 5$

26. Area of $\triangle\,PST$: area of $\triangle\,PQR =$

   A $1 : 2$

   B $1 : 4$

   C $2 : 3$

   D $4 : 9$

   E $3 : 2$

27. Area of $\triangle\,PST$: area of trapezium $STRQ =$

   A $4 : 5$

   B $5 : 4$

   C $1 : 3$

   D $1 : 1$

   E $2 : 1$

The table below applies to questions **28, 29** and **30**. Weekly time allocation for a certain group in a school.

Subjects	Arts	Sciences	Practical	P.E.
Time in hours	12	7	3	2

28. A pie chart is made from the table. The angle of the sector representing Sciences is

   A $15°$

   B $72 \cdot 5°$

   C $105°$

   D $165°$

   E $210°$

29. The percentage of time spent on Practical work is

   A 3%

   B 8%

   C $12\frac{1}{2}\%$

   D $43 \cdot 2\%$

   E 45%

30. A D.E.S. form requires to know what pupils are doing at a particular time chosen in the school week. The probability that this group will be doing a Science subject is

   A $\dfrac{7}{24}$

   B $\dfrac{7}{17}$

   C $\dfrac{11}{24}$

   D $\dfrac{11}{13}$

   E $\dfrac{7}{25}$

Test 17

Time allowed: 45 minutes

1. Which one of the following, when expressed to 3 significant figures, would be 4·35?

   A   0·435

   B   4·336

   C   4·344

   D   4·346

   E   4·356

2. When £60 is divided in the ratio 1 : 5 the numbers of £s in each part are

   A   12 and 48

   B   20 and 40

   C   16 and 54

   D   10 and 50

   E   15 and 45

3. The value of $3xy^2$ when $x = 3$, $y = -1$ is

   A   −81

   B   −9

   C   9

   D   18

   E   81

4. $P = \{a, b, c\}$, $Q = \{b, c, d, e, f\}$.
   $P \cup Q =$

   A $\{b, c\}$

   B $\{a, d, e, f\}$

   C $\{b, c, d, e, f\}$

   D $\{a, b, c, c, d, e, f\}$

   E $\{a, b, c, d, e, f\}$

5.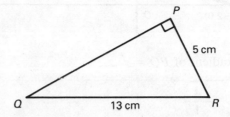

The area, in cm², of triangle PQR is

   A   32·5

   B   30

   C   60

   D   65

   E   78

6. Given that the circumference of a circle is 4 cm, then its radius, in cm, is

   A $\dfrac{2}{\pi}$

   B $\dfrac{4}{\pi}$

   C $\dfrac{8}{\pi}$

   D $\dfrac{2}{\sqrt{\pi}}$

   E $\sqrt{\dfrac{2}{\pi}}$

7. The price of an article after being increased by a tax of 10% is £33. The price before tax was added was

   A   £3

   B   £3.30

   C   £29.70

   D   £30

   E   £36.30

8.

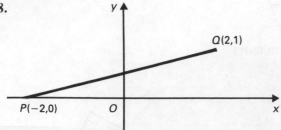

Gradient of $PQ =$

A $-\dfrac{4}{1}$

B $-\dfrac{1}{4}$

C $\dfrac{1}{4}$

D $\dfrac{1}{2}$

E $\dfrac{4}{1}$

9.

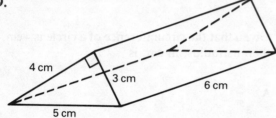

The volume, in cm^3, of the prism shown is

A   12

B   36

C   72

D   90

E   120

10. 5% of 6% of £120 is

A   £13.20

B   £1.32

C   36p

D   £3.60

E   £36

11. 0·024 expressed in standard form is

A $2·4 \times 10^2$

B $0·24 \times 10^{-1}$

C 24×10^{-3}

D $2·4 \times 10^{-2}$

E 24×10^3

12. The mean and mode of the numbers 2, 2, 2, 2, 4, 4, 5, 9, 15 are, respectively

A   2, 5

B   5, 2

C   5, 4

D   5, 15

E   2, 4

13.

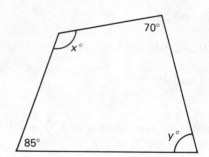

$x + y =$

A   25

B   105

C   155

D   180

E   205

14. The points (1, 0) and (2, 0) both lie on the line

A $x = 0$

B $y = 0$

C $x = 3$

D $x + y = 3$

E $x - y = 1$

15. In a shop Maria spent £25.80 plus VAT at 15%. Her total bill was

A   £3.87

B   £29.67

C   £30.67

D   £38.70

E   £40.80

16. A bag contains just 2 red, 3 white and 7 black balls. A ball is to be drawn at random. The probability that it will *not* be black is

A $\dfrac{1}{3}$

B $\dfrac{2}{3}$

C $\dfrac{5}{12}$

D $\dfrac{7}{12}$

E $\dfrac{5}{7}$

17. $2x + y = 5$,
$3x - y = 10$.

To satisfy both equations simultaneously, $x =$

A -5

B -1

C 1

D 3

E 5

18. The angles of a triangle are $(x + 20)°$, $(2x - 40)°$ and $(x + 50)°$.
$x =$

A   10

B $11\frac{2}{3}$

C   25

D   30

E   55

19.

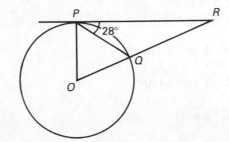

O is the centre of the circle and PR is a tangent.
$\angle OQP =$

A   72°

B   62°

C   59°

D   56°

E   28°

20. $v = u + ft$,
$f =$

A $\dfrac{v - u}{t}$

B $\dfrac{u - v}{t}$

C $\dfrac{vu}{t}$

D $\dfrac{t}{v - u}$

E $\dfrac{t}{u - v}$

21.

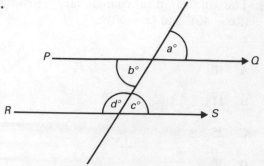

PQ is parallel to *RS*.
Which of the following is (are) true?
1 $a = b$
2 $a = c$
3 $c + d = 90$

A   **1** and **2** only

B   **1** and **3** only

C   **1** only

D   **2** only

E   all of them

22. A man cycles at a steady speed of 18 km/h. In 1 second he covers

A   500 m

B   300 m

C   50 m

D   30 m

E   5 m

23. A girl exchanged £50 for 132 dollars. The number of dollars she would have obtained for £15 at the same rate is

A   440

B   396

C   39·6

D   39·3

E   26·4

24. $\mathscr{E}$ = {quadrilaterals}, P = {parallelograms}, R = {rectangles}, S = {squares}.
Which one of the following is true?

A $P \cup R = \mathscr{E}$

B $P \cap R = S$

C $R \subset S$

D $R' \cap P = \varnothing$

E $R \subset P$

25.

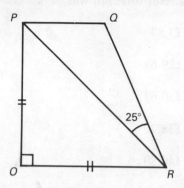

The figure is in a vertical plane with *OR* and *PQ* horizontal and *OP* = *OR*.
The angle of depression of *R* from *Q* is

A   20°

B   50°

C   60°

D   65°

E   70°

The table below applies to questions **26, 27** and **28**.

Value of x	1	2	3	4	5
Frequency of x	1	2	3	4	5

26. The mean value of x is

A $2\frac{1}{2}$

B $3\frac{2}{3}$

C 3

D 4

E 5

27. The mode of the distribution

A is 5

B is 4

C is $3\frac{2}{3}$

D is 3

E is 1, 2, 3, 4 or 5

28. The median of the distribution is

A 5

B $4\frac{1}{2}$

C 4

D $3\frac{2}{3}$

E 3

The diagram, which shows a cylindrical roller, applies to questions **29** and **30**.
Take $\pi = 3 \cdot 14$ in these questions.

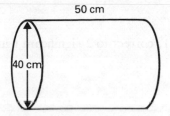

29. The roller is made of thin metal and is given weight by filling with water. The volume of water, in cm³ to 2 significant figures, is

A 6300

B 13 000

C 31 000

D 63 000

E 250 000

30. The area, in cm² to 2 significant figures, on the ground covered by one complete turn of the roller is

A 6300

B 63 000

C 3100

D 31 000

E 2000

73

Test 18

Time allowed: 45 minutes

1. 1570 ÷ 11, correct to 2 significant figures, is

A 150

B 142·73

C 142·72

D 140

E 14

2. $\frac{1}{2}$ of $(2\frac{1}{3} - 1\frac{1}{6}) =$

A $\frac{7}{12}$

B $\frac{2}{3}$

C $1\frac{1}{12}$

D $1\frac{3}{4}$

E $2\frac{1}{3}$

3. $3\frac{1}{8} - 2\frac{1}{6} =$

A $\frac{1}{24}$

B $\frac{23}{24}$

C $1\frac{1}{24}$

D $1\frac{1}{2}$

E $1\frac{23}{24}$

4. Given that the bearing of P from Q is 210°, then the bearing of Q from P is

A 210°

B 150°

C 120°

D 060°

E 030°

5. $\mathscr{E}$ = {positive integers}; P = {prime numbers greater than 2}; E = {positive even numbers}; O = {positive odd numbers}.
Which one of the following statements is *not* true?

A $E \cup O = \mathscr{E}$

B $P \subset O$

C $E \cap O = \varnothing$

D $P \cap E = \varnothing$

E $P \cup E = \mathscr{E}$

6. $(0·06)^2 =$

A 0·0036

B 0·012

C 0·036

D 0·12

E 0·36

7. A video recorder is priced at £416 when paid for by a 25% deposit, the remainder being paid in 12 monthly instalments. The monthly instalment is

A £8.67 to nearest p

B £26 exactly

C £34.67 to nearest p

C £40.83 to nearest p

E £104 exactly

74

8. The solutions of the simultaneous equations

$$x - 2y = 3,$$
$$x + 2y = 11$$

are

A $x = 7, y = 2$

B $x = 2, y = 7$

C $x = 4, y = 3\frac{1}{2}$

D $x = -1, y = -2$

E $x = 7, y = -2$

9. When $x = -2$, $(x - 5)^2 =$

A -49

B -7

C 9

D 29

E 49

10.

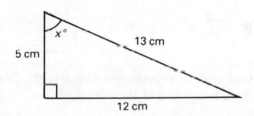

$\tan x° =$

A $\dfrac{12}{13}$

B $\dfrac{12}{5}$

C $\dfrac{5}{12}$

D $\dfrac{13}{12}$

E $\dfrac{5}{13}$

11. 5 kg is divided into 4 parts in the ratios $1:2:3:4$. The smallest part, in g, is

A 50

B 250

C 500

D 1250

E 2000

12. The exterior angle of a regular 12-sided polygon is

A $150°$

B $75°$

C $60°$

D $30°$

E $15°$

13. The volume, in cm^3, of a circular cylinder of height 4 cm and base radius 3 cm is

A 9π

B 12π

C 24π

D 36π

E 48π

14. The probability that, in two throws of a die, the total score will be six is

A $\dfrac{1}{36}$

B $\dfrac{5}{36}$

C $\dfrac{6}{36}$

D $\dfrac{5}{12}$

E $\dfrac{6}{12}$

15.

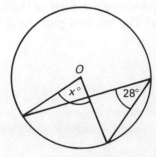

O is the centre of the circle. $x =$

A 14

B 28

C 56

D 62

E 152

16. A rectangular forest has length 2·3 km and breadth 1·3 km. Correct to the nearest km², its area is

A 1

B 2

C 3

D 4

E 5

17. A 15 m length of braid is cut into two pieces. One piece is ⅔ the length of the other. The length, in m, of the longer piece is

A 12

B 10

C 9

D 6

E 5

18. Bankim swings through an angle of 120° on a garden swing of which the chains are 2 m long. The total distance he covers, in m, on each forward swing is

A 4π

B 2π

C $\dfrac{4\pi}{3}$

D $\dfrac{2\pi}{3}$

E $\dfrac{\pi}{3}$

19. The gradient of the line joining the point (1, 2) to the point (3, 7) is

A $\dfrac{5}{2}$

B $\dfrac{9}{4}$

C $\dfrac{2}{5}$

D $\dfrac{-2}{5}$

E $\dfrac{-5}{2}$

20. One angle of an isosceles triangle is 42°. One of the other angles of the triangle

A is 42°

B is 138°

C is 69°

D is 64°

E cannot be determined without further information

21.

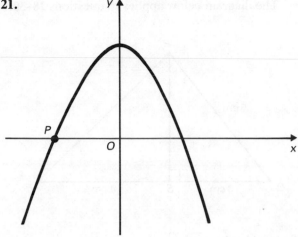

The diagram is part of the graph of the curve $y = 4 - x^2$. The point P is

A $(0, 4)$

B $(0, 2)$

C $(0, -2)$

D $(2, 0)$

E $(-2, 0)$

22.

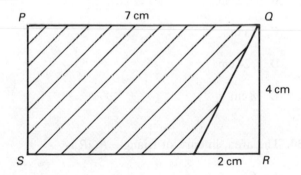

$PQRS$ is a rectangle. The area, in cm^2, of the shaded region is

A 48

B 28

C 24

D 20

E 8

23. Given that x is a positive integer, the least value of x for which $2^x > 18$ is

A 4

B 5

C 9

D 10

E 32

24. A man borrowed £x on the 1st January on which interest of 15% is charged each year. The total amount he owes at the end of 1 year is

A £$\dfrac{3x}{20}$

B £$\dfrac{4x}{20}$

C £$\dfrac{17x}{20}$

D £$\dfrac{23x}{20}$

E £ $16x$

25. In a school year group of 180 pupils there is a choice between Art, Woodwork, Metalwork and Home Economics. 98 pupils choose Art. In a pie chart the angle of the sector representing Art would be

A 49°

B 90°

C 98°

D 182°

E 196°

26. A cricketer scores 34, 36, 30, 44 in four successive innings. What must he score in the fifth innings if the mean of the five scores is 35?

A 31

B 34

C 35

D 39

E 144

27. The formula $P = 2\pi r + 2\ell$, when rewritten with r as the subject becomes $r =$

A $P - 2\ell$

B $\frac{1}{2}P - \ell$

C $\dfrac{2\ell - P}{2\pi}$

D $\dfrac{P - 2\ell}{2\pi}$

E $\dfrac{P + 2\ell}{2\pi}$

The diagram below applies to questions **28–30**.

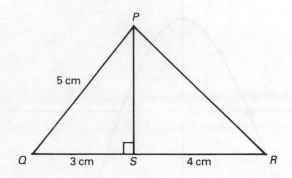

28. $\tan Q =$

A $\frac{3}{5}$

B $\frac{4}{5}$

C $\frac{3}{4}$

D $\frac{4}{3}$

E $1\frac{1}{4}$

29. $PR =$

A 4 cm

B $4\sqrt{2}$ cm

C 5 cm

D $\sqrt{24}$ cm

E 8 cm

30. The area, in cm², of triangle PQR is

A 6

B 8

C 12

D 14

E 28

78

Test 19

Time allowed: 45 minutes

1. 6% of £2400 is

A   £400

B   £144

C   £14.40

D   £4

E   £1.44

2. To make a cake for 6 people 150 g of butter is needed. The weight, in g, needed for a similar cake for 10 people is

A   25

B   90

C   200

D   250

E   500

3. Given that $R = \{1, 2\}$, $S = \{3, 4\}$ and $T = \{4\}$, then
$R \cup (S \cap T) =$

A $\{1, 2, 3, 4\}$

B $\{1, 2, 4\}$

C $\{1, 2, 3\}$

D $\{4\}$

E $\varnothing$

4. A triangle with base 12 cm has area 60 cm^2. Its height, in cm, is

A $2\frac{1}{2}$

B   5

C   10

D   12

E   20

5. $O = \{$odd numbers$\}$, $E = \{$even numbers$\}$, $x \in O$, $y \in E$. Which *one* of the following statements is *false*?

A $x^2 \in O$

B $y^2 \in E$

C $3y \in E$

D $xy \in O$

E $3x \in O$

6. Given that $\mathbf{p} = \begin{pmatrix} 2 \\ -1 \end{pmatrix}$, $\mathbf{q} = \begin{pmatrix} 1 \\ -2 \end{pmatrix}$ and $\mathbf{r} + \mathbf{p} = 2\mathbf{q}$, then $\mathbf{r} =$

A $\begin{pmatrix} 0 \\ -5 \end{pmatrix}$

B $\begin{pmatrix} 0 \\ 5 \end{pmatrix}$

C $\begin{pmatrix} 0 \\ -3 \end{pmatrix}$

D $\begin{pmatrix} 4 \\ -3 \end{pmatrix}$

E $\begin{pmatrix} 4 \\ -5 \end{pmatrix}$

7. The total rateable value of a borough is £50 000 000. The product of a 1p rate is

A   £5000

B   £50 000

C   £500 000

D   £5 000 000

E   £50 000 000

8. Given that $\dfrac{x}{2} + \dfrac{1}{4} = 1$, then $x =$

  A $\frac{3}{8}$

  B $\frac{3}{2}$

  C $\frac{5}{2}$

  D 5

  E 7

9. Which one of the following points does *not* lie on the curve $y = x^2 - 1$?

  A $(-1, 2)$

  B $(0, -1)$

  C $(1, 0)$

  D $(-2, 3)$

  E $(-10, 99)$

10. Given that $\mathbf{M} = \begin{pmatrix} 4 & -2 \\ 3 & -1 \end{pmatrix}$, the inverse $\mathbf{M}^{-1} =$

  A $\begin{pmatrix} -1 & 2 \\ -3 & 4 \end{pmatrix}$

  B $\begin{pmatrix} -1 & -3 \\ 2 & 4 \end{pmatrix}$

  C $\begin{pmatrix} -\frac{1}{2} & -1\frac{1}{2} \\ 1 & 2 \end{pmatrix}$

  D $\begin{pmatrix} -2 & 1\frac{1}{2} \\ -1 & \frac{1}{2} \end{pmatrix}$

  E $\begin{pmatrix} -\frac{1}{2} & 1 \\ -1\frac{1}{2} & 2 \end{pmatrix}$

11. The length, in km, represented by 1 cm on a map whose scale is 1 : 10 000 is

  A 0·1

  B 1

  C 10

  D 100

  E 1000

12. $(x - 3)$ is a factor of $x^2 - px + 21$.
$p =$

  A -10

  B 4

  C -4

  D 7

  E 10

13.

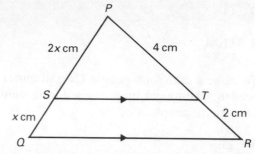

ST is parallel to QR. What is the value of x?

  A 1

  B 2

  C 3

  D 4

  E It cannot be found

14.

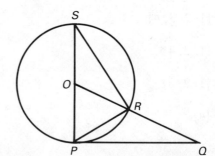

O is the centre of the circle and PQ is a tangent. Which *one* of the following is *not always* true?

  A SP is perpendicular to PQ

  B $OR = RP$

  C SR is perpendicular to PR

  D $\angle RPQ = \angle PSR$

  E $OR = OP$

15. A wire stay of length 5 m is attached to a point 4 m up a vertical pole, the other end being fastened to horizontal ground through the foot of the pole. The inclination, $\theta°$, of the wire stay to the horizontal is given by

A $\sin \theta° = \frac{4}{5}$

B $\sin \theta° = \frac{5}{4}$

C $\tan \theta° = \frac{3}{4}$

D $\cos \theta° = \frac{4}{5}$

E $\cos \theta° = \frac{5}{4}$

16. $f : x \mapsto x^3 + 3x^2 - 2x + 1$.
Which of the following is (are) true?
1 $f(1) = 3$
2 $f(-1) = 5$
3 $f(0) = 0$

A   **1** only

B   **2** only

C   **3** only

D   **1** and **2** only

E   **1, 2** and **3**

17.

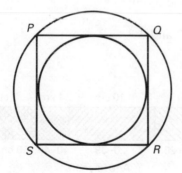

PQRS is a square. The ratio

   area of inner circle : area of outer circle

A $= \pi/4$

B $= 2 : 1$

C $= 1 : \sqrt{2}$

D $= 1 : 2$

E   cannot be found

18.

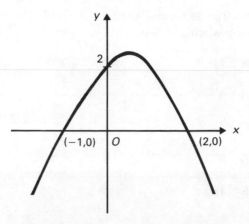

The diagram could be a sketch graph of

A $y = (x + 1)(x - 2)$

B $y = (x + 1)(x + 2)$

C $y = (1 - x)(2 + x)$

D $y = (1 - x)(2 - x)$

E $y = (1 + x)(2 - x)$

19. When the subject of the formula $k = \dfrac{t + 1}{t}$ is changed to t, it becomes

$t =$

A $k - 1$

B $\dfrac{1}{k}$

C $\dfrac{1}{k - 1}$

D $\dfrac{1}{k + 1}$

E $k\,t - 1$

20. A quadratic equation whose roots are $-\frac{1}{2}$ and $\frac{2}{3}$ is

A $(2x - 1)(2x - 3) = 0$

B $(2x - 1)(3x - 2) = 0$

C $(2x + 1)(2x + 3) = 0$

D $(x + 2)(2x - 3) = 0$

E $(2x + 1)(3x - 2) = 0$

21. y varies as x^2, and when $x = 2$, $y = 2$. The value of y when $x = 3$ is

A $1\frac{1}{2}$

B 3

C $4\frac{1}{2}$

D 9

E 18

22. Three fair dice are thrown. The probability that three even numbers will show is

A $\dfrac{1}{2}$

B $\dfrac{1}{3}$

C $\dfrac{1}{6}$

D $\dfrac{1}{8}$

E none of these

23.

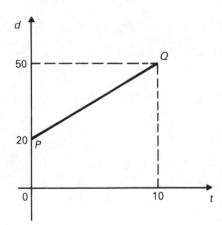

The diagram shows the time–distance graph PQ of a journey, where distance d is in metres and time t in seconds. The speed, in m/s, for the journey is

A $\frac{1}{3}$

B 2

C 3

D 3·5

E 5

24. When x is an integer, the least value of x which satisfies the inequality

$$5 - 6x < 2 - 4x$$

is

A 2

B 1

C 0

D −1

E −2

25. For which of the following transformation matrices is the origin an invariant point?

1 $\begin{pmatrix} 2 & 0 \\ 0 & 2 \end{pmatrix}$

2 $\begin{pmatrix} 0 & -1 \\ 1 & 0 \end{pmatrix}$

3 $\begin{pmatrix} 2 & 1 \\ 1 & 2 \end{pmatrix}$

A **1** only

B **1** and **2** only

C **1** and **3** only

D **2** and **3** only

E **1, 2** and **3**

26.

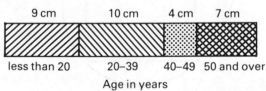

The component bar chart is of total length 30 cm and shows the distribution of ages of 15 000 people. How many of the people are under 50 years old?

A 2000

B 3500

C 7000

D 9500

E 11 500

27. An enlargement has centre (1, 0) and scale factor 3. The image of the point *P* (2, 3) under this enlargement is *P'*. The coordinates of *P'* are

A (6, 9)

B (4, 9)

C (5, 9)

D (5, 12)

E (3, 6)

28. A quadrilateral is symmetrical about one of its diagonals. It follows that the quadrilateral *must* be a

A trapezium

B rectangle

C rhombus

D kite

E square

29. Given that $Q \cap P' = \varnothing$ and that $P \neq Q$, then

A $P \cup Q = P$

B $P \cap Q' = P$

C $Q \cup P' = \mathscr{E}$

D $P \cap Q = Q'$

E $P \cup Q' = P'$

30.

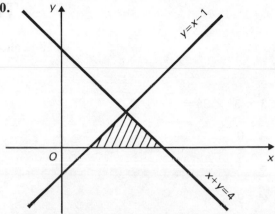

The coordinates of any point (*x*, *y*) which lies within the shaded region shown in the diagram must satisfy the following three inequalities

A $y > x - 1, x + y < 4, x > 0$

B $y < x - 1, x + y < 4, y > 0$

C $y < x - 1, x + y > 4, y > 0$

D $y > x - 1, x + y > 4, x > 0$

E $y < x - 1, x + y < 4, y < 0$

Test 20

Time allowed: 45 minutes

1. $3 \times 3^0 \times 3^2 =$

A 0

B 3

C 9

D 27

E 81

2. $\dfrac{1}{4x} + \dfrac{x}{3} =$

A $\dfrac{3 + 4x^2}{12x}$

B $\dfrac{1 + x}{4x + 3}$

C $\dfrac{3 + 4x}{12}$

D $\dfrac{4 + x}{4}$

E $\dfrac{3 + 4x}{12x}$

3.

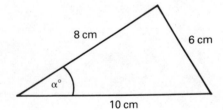

$\tan \alpha° =$

A 0·8

B 0·6

C 0·75

D 1·$\dot{3}$

E 1·25

4. Given that $p = -2$ and $q = 3$,
then $p^2 - pq =$

A -10

B -2

C 2

D 9

E 10

5. $\begin{pmatrix} 2 & 3 \\ 4 & 5 \end{pmatrix} \begin{pmatrix} 2 \\ 1 \end{pmatrix} =$

A $\begin{pmatrix} 7 \\ 13 \end{pmatrix}$

B $\begin{pmatrix} 10 \\ 9 \end{pmatrix}$

C $(2 \quad 1)$

D $(1 \quad 2)$

E $\begin{pmatrix} 4 & 6 \\ 4 & 5 \end{pmatrix}$

6. $6x^2 + x - 1 =$

A $(3x + 1)(2x - 1)$

B $(3x - 1)(2x + 1)$

C $(3x - 1)(2x - 1)$

D $(6x + 1)(x - 1)$

E $(6x - 1)(x + 1)$

7. The transformation represented by the matrix
$\begin{pmatrix} 2 & 0 \\ 0 & -2 \end{pmatrix}$ is

A a reflection

B a rotation

C a translation

D an enlargement

E none of these

8. Two similar jugs in a set are such that the height of the larger is twice that of the smaller. The volume of the larger is k times that of the smaller, where $k =$

A 2

B 3

C 4

D 6

E 8

9. The price of a second-hand car is reduced in a sale from £6000 to £5400. The reduction, expressed as a percentage of the original price, is

A 1%

B 6%

C 10%

D $11\frac{1}{9}$%

E 90%

10. Given that $3 : x = x : 12$, where $x > 0$, then $x =$

A 18

B 9

C $7\frac{1}{2}$

D 6

E 4

11. $0{\cdot}004\,48 \div 0{\cdot}32$ written in standard form is

A $1{\cdot}4 \times 10^{-1}$

B $1{\cdot}4 \times 10^{-3}$

C $1{\cdot}4 \times 10^{1}$

D $1{\cdot}4 \times 10^{-2}$

E $1{\cdot}4 \times 10^{0}$

12.

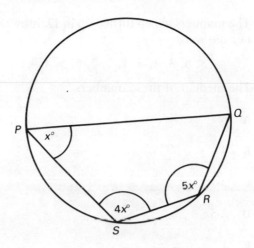

$x =$

A 30

B 60

C 90

D 120

E cannot be found

13.

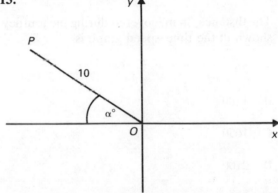

Given that $\cos \alpha° = \dfrac{3}{5}$ and $OP = 10$ units, the coordinates of P are

A $(-8, 6)$

B $(-6, 8)$

C $(6, -8)$

D $(8, -6)$

E $(-6, -6)$

85

14. The numbers which turned up in 12 throws of a fair die were

$$5, 6, 3, 2, 6, 1, 1, 2, 6, 5, 4, 4.$$

The median of these numbers is

A 6

B 4

C 3·75

D 3·5

E 1

15.

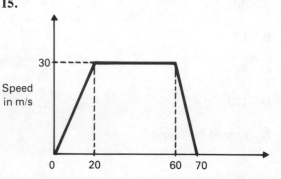

The distance, in m, covered during the journey shown in the time–speed graph is

A 30

B 1200

C 1650

D 2100

E 3300

16. The solutions of the quadratic equation $x^2 - 3x - 1 = 0$ are

A $\dfrac{3 \pm \sqrt{13}}{2}$

B $\dfrac{-3 \pm \sqrt{13}}{2}$

C $\dfrac{3 \pm \sqrt{5}}{2}$

D $\dfrac{-3 \pm \sqrt{5}}{2}$

E $\dfrac{3 \pm \sqrt{7}}{2}$

17.

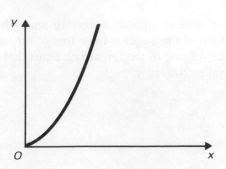

Which one of the following relations for positive values of x and y could the given graph represent?

A y is directly proportional to x

B y is directly proportional to x^2

C y is inversely proportional to x

D y^2 is inversely proportional to x

E y is directly proportional to $\sqrt{x}$

18. $f : x \mapsto x^2 + 2$ for $x \in \mathbb{R}$ and $g : x \mapsto 2x + 1$ for $x \in \mathbb{R}$.
Then, for $x \in \mathbb{R}$, $fg : x \mapsto$

A $2x^2 + 5$

B $2x^2 + 3$

C $2x^3 + x^2 + 4x + 2$

D $4x^2 + 4x + 9$

E $4x^2 + 4x + 3$

19. The internal dimensions of a rectangular box are $6\,\text{cm} \times 8\,\text{cm} \times 24\,\text{cm}$. The length, in cm, of the longest diagonal of the box is

A 24

B 26

C 28

D 30

E 32

20. Given that x and y are positive integers and that $3x + y < 8$, then the number of points (x, y) satisfying these conditions is

A   1

B   2

C   3

D   4

E   5

21. The cost of 30 textbooks is £33 and the cost of 20 rulers is £6.60. The ratio of the cost of a textbook to the cost of a ruler is

A   5:1

B   1:3

C   3:10

D   10:3

E   3:2

22.

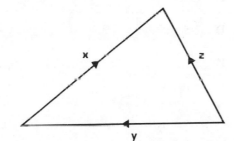

z =

A   x − y

B   y − x

C   x + y

D   ½(x + y)

E   −x − y

23. From an observation point O on the top of a tower of height H metres, the angle of depression of a point P on horizontal ground is $\theta°$. The horizontal distance from P to the foot of the tower is

A $\dfrac{H}{\tan \theta°}$

B $H \tan \theta°$

C $\dfrac{H}{\sin \theta°}$

D $H \sin \theta°$

E $H \cos \theta°$

24. The x-coordinates of the points of intersection of the curves $y = x^2$ and $3y = 4x - 1$ are given by the roots of the equation

A $x^2 = 4x - 1$

B $x^2 = \dfrac{4x}{3} - 1$

C $\dfrac{x^2}{3} = 4x - 1$

D $3x^2 = (4x - 1)^2$

E $3x^2 = 4x - 1$

25. A solid cylinder, of radius 6 cm and height 4 cm weighs 8 kg. The weight, in kg, of a cylinder made of the same material and of radius 3 cm and height 8 cm is

A   2

B   4

C   8

D   32

E   64

26. $P(X) = \dfrac{3}{4}$ and $P(Y) = \dfrac{1}{5}$, where X and Y are independent events.

$P(X \cap Y) =$

A $\dfrac{19}{20}$

B $\dfrac{11}{20}$

C $\dfrac{4}{15}$

D $\dfrac{3}{20}$

E $\dfrac{1}{20}$

27. An arc of a circle of radius 12 cm subtends an angle of 45° at the centre of the circle. The length, in cm, of the arc is

A 9π

B 6π

C $\dfrac{9\pi}{2}$

D 3π

E $\dfrac{3\pi}{2}$

28.

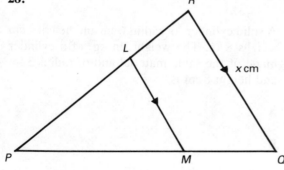

In the figure LM is parallel to RQ and $\dfrac{PM}{MQ} = \dfrac{3}{2}$. Then $LM =$

A $\dfrac{3x}{2}$

B $\dfrac{3x}{5}$

C $\dfrac{2x}{5}$

D $\dfrac{2x}{3}$

E $\dfrac{3}{5x}$

29. A speed of 72 km/h is the same as a speed of

A 2 m/s

B 12 m/s

C 20 m/s

D 120 m/s

E 200 m/s

30. A boy kept a check of the number of people in each car entering a car park. He checked 50 cars with the following results

No. of people per car	1	2	3	4
No. of cars	5	30	10	5

The mean number of people per car was

A 2

B 2·03

C 2·3

D 2·5

E 3

Test 21

1. $2^2 \times 2^3 \times 2^x = 2^{12}$. Then $x =$

   A   2

   B   3

   C   4

   D   6

   E   7

2. The cost of 50 g of silver at £500 per kg is

   A   £2.50

   B   £10

   C   £25

   D   £250

   E   £2500

3. The following numbers of people taking holidays in Spain, Greece and Portugal are represented in a pie chart.

Spain	44 000
Greece	16 000
Portugal	12 000

   The angle of the sector representing numbers of people going to Greece is

   A   160°

   B   80°

   C   32°

   D   16°

   E   8°

4. The graph of $y = x^2 - 2x - 15$ cuts the x-axis at two points whose distance apart is

   A   2 units

   B   8 units

   C   12 units

   D   14 units

   E   15 units

5. $\mathbf{M} = \begin{pmatrix} 0 & 5 \\ 5 & 0 \end{pmatrix}$. Then $\mathbf{M}^2 =$

   A $\begin{pmatrix} 25 & 0 \\ 0 & 25 \end{pmatrix}$

   B $\begin{pmatrix} 25 & 10 \\ 10 & 25 \end{pmatrix}$

   C $\begin{pmatrix} 0 & 25 \\ 25 & 0 \end{pmatrix}$

   D $\begin{pmatrix} 0 & 10 \\ 10 & 0 \end{pmatrix}$

   E   none of these

6.

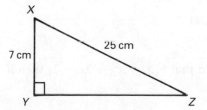

   The area, in cm^2, of $\triangle XYZ$ is

   A   84

   B $87\frac{1}{2}$

   C   168

   D   175

   E   300

7. The ratio of the length of a photo to that of its enlargement is $1:4$. The ratio of the area of the photo to that of its enlargement is

A $1:2$

B $1:4$

C $1:8$

D $1:16$

E $1:64$

8.

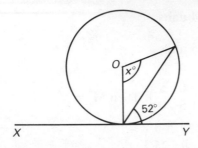

XY is a tangent to the circle, centre O.
Then $x =$

A   26

B   38

C   52

D   76

E   104

9. Given that $f:x \mapsto x^2 - 3x - 2$, then $f(-2)$ is

A   8

B   4

C   0

D -4

E -12

10. A point $(1, -2)$ is reflected in the y-axis and this image is reflected in the line $y = x$. The resulting point is

A $(-1, -2)$

B $(-1, 2)$

C $(-2, -1)$

D $(2, 1)$

E $(1, 2)$

11. Given that n(P) = 18, n(Q) = 24 and n($P \cap Q$) = 5, then n($P \cup Q$) =

A   5

B   32

C   37

D   42

E   none of these

12. The solution of the equation

$$\frac{3}{3x + 2} = \frac{1}{2}$$

is $x =$

A $-\frac{1}{6}$

B   0

C $\frac{1}{3}$

D $1\frac{1}{3}$

E $2\frac{1}{3}$

13. The modulus of the vector $\begin{pmatrix} -8 \\ 15 \end{pmatrix}$ is

A -120

B -7

C   17

D   7

E   289

90

14.

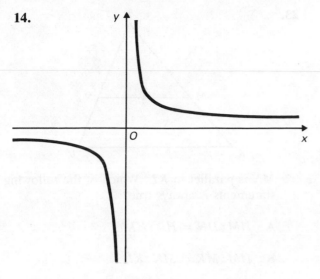

The equation of the curve shown in the diagram could be

A $\quad y = \dfrac{1}{x^2}$

B $\quad y = \dfrac{1}{x}$

C $\quad y = \dfrac{1}{x^4}$

D $\quad y = \dfrac{1}{1-x}$

E $\quad y = x^3$

15. The inverse of the matrix $\begin{pmatrix} 2 & 3 \\ 4 & x \end{pmatrix}$ cannot be found when $x =$

A $\quad -6$

B $\quad 1\frac{1}{2}$

C $\quad 0$

D $\quad 5$

E $\quad 6$

16. Given that $f : x \mapsto x^2 - 5$, with domain $-3 \leqslant x \leqslant 3$, the range of f is

A $\quad 0 \leqslant f(x) \leqslant 9$

B $\quad -5 \leqslant f(x) \leqslant 4$

C $\quad -5 \leqslant f(x) \leqslant 9$

D $\quad -5 \leqslant f(x) \leqslant 9$

E $\quad 0 \leqslant f(x) \leqslant 4$

17. Given that your mean mark for English and French is 50% and your mean mark for Maths, Chemistry and Physics is 70%, then your mean mark for all five subjects is

A $\quad 24\%$

B $\quad 48\frac{1}{3}\%$

C $\quad 58\%$

D $\quad 60\%$

E $\quad 62\%$

18.

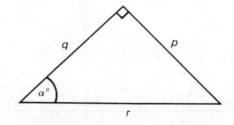

$r =$

A $\quad p \sin \alpha°$

B $\quad \dfrac{p}{\tan \alpha°}$

C $\quad q \cos \alpha°$

D $\quad \dfrac{p}{\sin \alpha°}$

E $\quad \dfrac{q}{\sin \alpha°}$

19. The solution set of the equation
$$2x^2 + x - 3 = 0$$
is

A $\quad \{-1\frac{1}{2},\ 1\}$

B $\quad \{1\frac{1}{2},\ -1\}$

C $\quad \{1\frac{1}{2},\ 1\}$

D $\quad \{\frac{1}{2},\ -3\}$

E $\quad \{-\frac{1}{2},\ 3\}$

20. The straight line joining the two points $(-2, 3)$ and $(1, 5)$ has a gradient equal to

A $\dfrac{3}{2}$

B $\dfrac{2}{3}$

C $-\dfrac{1}{8}$

D -2

E -8

21.

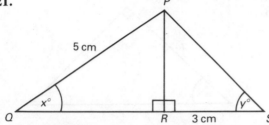

Given that $\sin x° = \dfrac{4}{5}$, then $\cos y° =$

A $\dfrac{4}{5}$

B $\dfrac{5}{4}$

C $\dfrac{5}{3}$

D $\dfrac{3}{5}$

E $\dfrac{3}{4}$

22. Three dice are to be thrown. The probability that the number which shows on each die will be greater than 4 is

A $\dfrac{1}{27}$

B $\dfrac{1}{9}$

C $\dfrac{1}{8}$

D $\dfrac{1}{3}$

E 1

23.

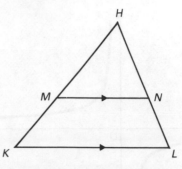

MN is parallel to KL. Which of the following statements is always true?

A $HM : HN = HL : HK$

B $HM : MK = MN : KL$

C $MN = \frac{1}{2}KL$

D $HK : HM = MN : KL$

E $HM : HK = HN : HL$

24. Which one of the following sets shown on the real number line is the solution set of the inequality $4x - 8 \leqslant 4 - 2x$?

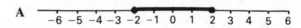

25. The transformation represented by the matrix $\begin{pmatrix} -2 & 0 \\ 0 & -2 \end{pmatrix}$ is

A a rotation

B a reflection

C an enlargement

D a translation

E none of these

26. The circumference of a cylindrical garden roller is 90 cm and the width is 50 cm. The area, in cm², of ground covered when the roller makes 10 complete revolutions is

A $\pi \times 50^2 \times 10$

B $\pi \times 90^2 \times 50 \times 10$

C   45 000

D   4500

E   500

27.

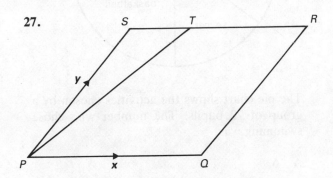

$PQRS$ is a parallelogram and $ST : TR = 1 : 2$.

Given that $\overrightarrow{PQ} = \mathbf{x}$ and $\overrightarrow{PS} = \mathbf{y}$, then $\overrightarrow{PT} =$

A $\mathbf{x} + \mathbf{y}$

B $\mathbf{y} + \frac{1}{2}\mathbf{x}$

C $\mathbf{y} - \frac{1}{3}\mathbf{x}$

D $\mathbf{y} + \frac{1}{3}\mathbf{x}$

E $\mathbf{y} - \frac{1}{2}\mathbf{x}$

28. Given that n is a positive integer, which *one* of the following cannot be an even integer whatever the value of n?

A $n^2 - 1$

B $3n + 1$

C $n(2n + 1)$

D $2n - 1$

E $n(n - 1)$

29. Given that $\dfrac{1}{f} = \dfrac{1}{u} + \dfrac{1}{v}$, then $v =$

A $f - u$

B $\dfrac{u - f}{uf}$

C $\dfrac{uf}{u - f}$

D $\dfrac{uf}{f - u}$

E $-\dfrac{uf}{u + f}$

30. The light of a lighthouse is at 30 m above sea level. From an observer on the sea shore the angle of elevation of the light is 32°. The distance, in m, of the observer from the lighthouse is

A   30 tan 58°

B   30 tan 32°

C   30 cos 32°

D $\dfrac{30}{\tan 58°}$

E   30 sin 58°

Test 22

Time allowed: 1 hour

1. One element of the set $\{21, 31, 51, 81, 111\}$ is a prime number. It is

   A 21

   B 31

   C 51

   D 81

   E 111

2. When factorized $7x^2 - 11x - 6 =$

   A $(7x - 3)(x + 2)$

   B $(7x - 2)(x + 3)$

   C $(7x + 2)(x - 3)$

   D $(7x + 3)(x - 2)$

   E $(7x - 6)(x + 1)$

3. When $v = u + at$, then $a =$

   A $v - u - t$

   B $\dfrac{v - u}{t}$

   C $\dfrac{v}{t} - u$

   D $\dfrac{v + u}{t}$

   E $t(v - u)$

4. The population of Southland is $8\,624\,000$. This population, when expressed in standard form is

   A $0 \cdot 8624 \times 10^7$

   B $86 \cdot 24 \times 10^5$

   C $8 \cdot 624 \times 10^5$

   D $8 \cdot 624 \times 10^6$

   E $8 \cdot 624 \times 10^7$

5.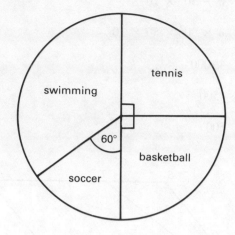

   The pie chart shows the activities chosen by a group of 72 pupils. The number who chose swimming was

   A 6

   B 12

   C 18

   D 24

   E 28

6. In a transformation with the 2×2 matrix **M** the vectors

   $\begin{pmatrix} 1 \\ 0 \end{pmatrix}$ and $\begin{pmatrix} 0 \\ 1 \end{pmatrix}$ are mapped onto the vectors

   $\begin{pmatrix} -2 \\ 3 \end{pmatrix}$ and $\begin{pmatrix} 3 \\ -4 \end{pmatrix}$ respectively.

   Then **M** =

   A $\begin{pmatrix} 3 & -2 \\ -4 & 3 \end{pmatrix}$

   B $\begin{pmatrix} 3 & -4 \\ -2 & 3 \end{pmatrix}$

   C $\begin{pmatrix} -2 & 3 \\ 3 & -4 \end{pmatrix}$

   D $\begin{pmatrix} -2 & -4 \\ 3 & 3 \end{pmatrix}$

   E $\begin{pmatrix} -2 & 3 \\ -4 & 3 \end{pmatrix}$

7.

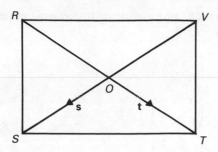

RSTV is a rectangle and $\overrightarrow{OS} = \mathbf{s}$, $\overrightarrow{OT} = \mathbf{t}$.
Then $\overrightarrow{VT} =$

A $\mathbf{s} - \mathbf{t}$

B $\mathbf{t} - \mathbf{s}$

C $\mathbf{s} + \mathbf{t}$

D $2\mathbf{s} - \mathbf{t}$

E $2\mathbf{s} + \mathbf{t}$

8. The mean weight of the 8 men rowing a racing boat is 87 kg. The weight of their cox is 51 kg. The mean weight, in kg, of the whole crew of 9 men is

A $93\frac{3}{8}$

B   83

C $82\frac{1}{2}$

D $80\frac{5}{8}$

E   69

9. The area under a time–speed graph gives a measure of

A   velocity

B   acceleration

C   average speed

D   average acceleration

E   distance covered

10.

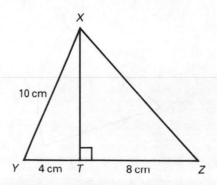

Measured in cm², $XZ^2 =$

A   148

B   180

C   196

D   208

E   244

11.

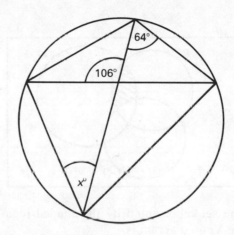

$x =$

A   21

B   32

C   42

D   53

E   64

12.

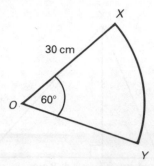

OXY is a sector of a circle, centre O and radius 30 cm. Taking $\pi = 3\cdot14$, the length, in cm, of arc XY is

A $15\cdot7$

B   30

C $31\cdot4$

D $62\cdot8$

E   471

13.

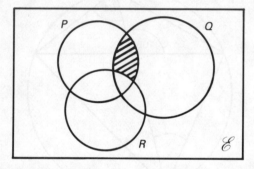

The set represented by the shaded region of the Venn diagram is

A $(P \cup Q) \cap R$

B $(P \cap Q) \cup R'$

C $(P \cap Q) \cap R$

D $P \cap (Q \cup R')$

E $(P \cap Q) \cap R'$

14. Which one of the following matrices is singular?

A $\begin{pmatrix} 2 & 3 \\ 3 & 5 \end{pmatrix}$

B $\begin{pmatrix} 3 & -2 \\ 6 & 4 \end{pmatrix}$

C $\begin{pmatrix} 3 & 6 \\ -4 & 8 \end{pmatrix}$

D $\begin{pmatrix} 2 & 4 \\ 5 & 10 \end{pmatrix}$

E $\begin{pmatrix} 2 & 0 \\ 0 & 1 \end{pmatrix}$

15.

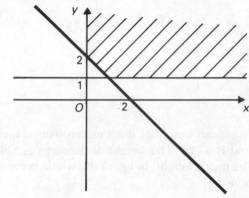

Any point (x, y) in the shaded region, not including its boundaries, must satisfy

A $x > 0, y > 1, x + y > 2$

B $x > 0, y > 1, x + y < 2$

C $x > 0, y < 1, x + y > 2$

D $x > 1, y > 1, y > x - 2$

E $x > 0, y < 2, y > 2 - x$

16. The number of points of intersection of the graphs of $y = 2x$ and $y = x^2$ is

A   0

B   1

C   2

D   3

E   4

17. $(x - 2y)^2 =$

A $x^2 - 4y^2$

B $x^2 + 4y^2$

C $x^2 - 2xy + 4y^2$

D $x^2 - 2xy + 2y^2$

E $x^2 - 4xy + 4y^2$

18.

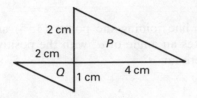

The mapping region $P \to$ region Q is an enlargement with scale factor

A -2

B $-\frac{1}{2}$

C 2

D $\frac{1}{2}$

E -1

19. Given that $3y = 2x$, the value of $\dfrac{y + x}{y - x}$ is

A 5

B 4

C 2

D -4

E -5

20. $\mathbf{p} = \begin{pmatrix} 2 \\ 3 \end{pmatrix}$ and $\mathbf{q} = \begin{pmatrix} 1 \\ 2 \end{pmatrix}$.

Which one of the following is equal to $\begin{pmatrix} 4 \\ 5 \end{pmatrix}$?

A $\mathbf{p} + 2\mathbf{q}$

B $4\mathbf{p} - 4\mathbf{q}$

C $\mathbf{p} + \mathbf{q}$

D $3\mathbf{p} - 2\mathbf{q}$

E $3\mathbf{q} - 2\mathbf{p}$

21. $f : x \mapsto 2x,\ g : x \mapsto x + 5$.
Then $fg : x \mapsto$

A $2x + 5$

B $2x + 10$

C $3x + 5$

D $2x^2 + 5x$

E $2x(x + 5)$

22.

x	1	-2	3	-4
y	12	-6	4	-3

Which one of the following could represent the relation between the values of x and y in the given table?

A y varies inversely as x

B y varies directly as x

C y varies inversely as x^2

D y varies directly as x^2

E y varies inversely as x^3

23. A cylinder and a sphere have the same radius r cm and the same volume. The height of the cylinder is

A $4r$

B $2r$

C $\frac{4}{3}$

D $\frac{1}{3}r$

E $\frac{4}{3}r$

24. Given that p and q are positive integers, then $\dfrac{p}{q}$ *must* be

A rational

B irrational

C an integer

D an odd number

E prime

25.

Number	10	8	6	4	2
Frequency	2	3	1	3	1

The mean m of the above distribution is such that

A $3 < m < 5$

B $5 < m < 6$

C $6 < m < 7$

D $7 < m < 8$

E $8 < m < 10$

26. $\dfrac{2}{x-2} - \dfrac{1}{x+2} =$

A $\dfrac{x+6}{x^2-4}$

B $\dfrac{3x+2}{x^2-4}$

C $\dfrac{x+4}{x^2-4}$

D $\dfrac{x}{x^2-4}$

E $\dfrac{1}{x-2}$

27. The solution set of the inequation $5 + x > 2 - 3x$ is

A $\{x : x > \frac{3}{4}\}$

B $\{x : x < \frac{3}{4}\}$

C $\{x : x < -\frac{3}{4}\}$

D $\{x : x > -\frac{3}{4}\}$

E $\varnothing$

28. The line joining the points $(3, 5)$ and $(2, 1)$ makes an angle of $\alpha°$ with the positive x-axis, where $\tan \alpha° =$

A $\dfrac{1}{4}$

B $\dfrac{1}{2}$

C -4

D $\dfrac{6}{5}$

E 4

29.

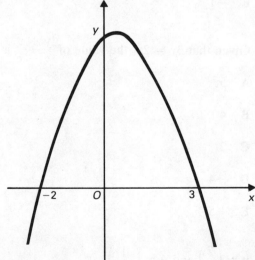

The diagram could be a sketch of part of the graph of $y =$

A $(x-3)(x+2)$

B $(x+3)(x-2)$

C $(x+2)(x+3)$

D $(2-x)(3+x)$

E $(2+x)(3-x)$

30. The probability of a train arriving on time is $\frac{3}{4}$. On 3 successive journeys, the probability that the train will be on time for the first two and arrive late for the third is

A $\dfrac{9}{64}$

B $\dfrac{9}{16}$

C $\dfrac{3}{64}$

D $\dfrac{3}{16}$

E $\dfrac{1}{3}$

31. When corrected to 3 significant figures, $0 \cdot 038087 =$

A $0 \cdot 038$

B $0 \cdot 03809$

C $0 \cdot 0381$

D $0 \cdot 040$

E $0 \cdot 04$

32. Which of the following is (arc) unit matrices?

1 $\begin{pmatrix} 1 & 1 \\ 1 & 1 \end{pmatrix}$

2 $\begin{pmatrix} 0 & 1 \\ 1 & 0 \end{pmatrix}$

3 $\begin{pmatrix} 1 & 0 \\ 0 & 1 \end{pmatrix}$

A 1 only

B 2 only

C 3 only

D 2 and 3 only

E 1, 2 and 3

33. $S = \{$positive multiples of 3$\}$, $T = \{$positive multiples of 5$\}$, $V = \{$positive multiples of 7$\}$. Then $(S \cup T) \cap V =$

A $\{$positive multiples of 21 or 35$\}$

B $\{$positive multiples of 105$\}$

B $\{$positive multiples of 15 or 35$\}$

D $\{$positive multiples of 15 or 21$\}$

E $\{$positive multiples of 70$\}$

34. The modulus of the vector

$$(9\mathbf{i} + 10\mathbf{j}) - (5\mathbf{i} + 7\mathbf{j})$$

is

A $\sqrt{505}$

B $\sqrt{217}$

C 7

D 5

E $3\frac{1}{2}$

35.

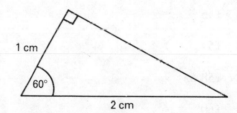

Which of the following is (are) rational?

1 sin 60°
2 cos 60°
3 tan 60°

A 1, 2 and 3

B 1 and 3 only

C 1 only

D 2 only

E 3 only

36. $8^{-\frac{1}{3}} \times 4^{\frac{1}{2}} =$

A $\frac{1}{2}$

B 1

C 2

D -2

E -4

37. Given that $x^2 + 6x - 1 \equiv (x + a)^2 + b$, then the constants a and b are given by

A $a = 3, b = -1$

B $a = -3, b = -1$

C $a = 3, b = 8$

D $a = -3, b = -10$

E $a = 3, b = -10$

38. A price is quoted as '£69 inclusive of VAT at 15%'. The price *before* the addition of VAT was

A £46

B £54

C £58.65

D £60

E £66

39. $\frac{1}{2}\%$ is the same as

1 0·05

2 $\dfrac{1}{200}$

3 5×10^{-3}

Which of the above is (are) correct?

A 1, 2 and 3

B 2 and 3 only

C 1 and 2 only

D 1 only

E 2 only

40. For which of the following functions is $f^{-1} = f$ (i.e. the function(s) is (are) self-inverse)?

1 $f : x \mapsto 3 - x$

2 $f : x \mapsto \dfrac{1}{x}$

3 $f : x \mapsto \dfrac{1}{1 + x}$

A 1 and 2 only

B 2 and 3 only

C 1 only

D 2 only

E 1, 2 and 3

Test 23

Time allowed: 1 hour

1.

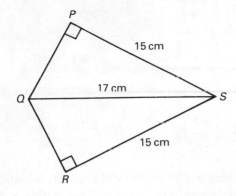

The area, in cm², of quadrilateral *PQRS*

A is 60

B is 68

C is 120

D is 136

E cannot be found from the information given

2. Given that $f : x \mapsto \dfrac{1}{x}$, then $f(3) - f(-3) =$

A 0

B $\frac{1}{3}$

C $\frac{2}{3}$

D 3

E 6

3. The smallest and greatest of the numbers $\dfrac{1}{7}$, $0 \cdot 14$, $14 \cdot 3 \times 10^{-2}$ are, respectively

A $0 \cdot 14$, $14 \cdot 3 \times 10^{-2}$

B $0 \cdot 14$, $\dfrac{1}{7}$

C $\dfrac{1}{7}$, $14 \cdot 3 \times 10^{-2}$

D $14 \cdot 3 \times 10^{-2}$, $\dfrac{1}{7}$

E $\dfrac{1}{7}$, $0 \cdot 14$

4.

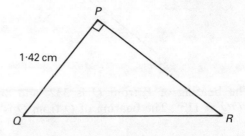

Given that $\tan \angle PRQ = \dfrac{2}{5}$ and $PQ = 1 \cdot 42$ cm, then the length, in cm to 2 significant figures, of *PR* is

A $3 \cdot 6$

B $3 \cdot 5$

C $0 \cdot 58$

D $0 \cdot 57$

E $0 \cdot 56$

5. $(8 \times 10^{10}) \div (4 \times 10^{5}) =$

A 2×10^{2}

B 4×10^{2}

C 4×10^{5}

D 2×10^{15}

E 2×10^{5}

6. Given that *y* varies directly as the square of *x*, then

A *x* varies directly as the square of *y*

B *x* varies directly as the square root of *y*

C *x* varies inversely as the square of *y*

D *x* varies inversely as the square root of *y*

E *x* varies inversely as *y*

7.

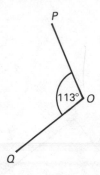

The bearing of P from O is 332° and angle $POQ = 113°$. The bearing of Q from O is

A   039°

B   085°

C   141°

D   219°

E   229°

8. Given that

$$3x + 2y = 7$$
$$2x + 3y = 3$$

which one of the following is *not* true?

A $x + y = 2$

B $x - y = 4$

C $x = 3$

D $y = 1$

E $y = -1$

9. The solution set of the equation
$$(x - 2)(x - 2) = 9$$
is

A $\{-\sqrt{5}, \sqrt{5}\}$

B $\{2\}$

C $\{5\}$

D $\{-5, 1\}$

E $\{5, -1\}$

10. $\begin{pmatrix} 0 & 1 \\ -1 & 0 \end{pmatrix}\begin{pmatrix} 3 \\ -5 \end{pmatrix} =$

A $\begin{pmatrix} 3 \\ -5 \end{pmatrix}$

B $\begin{pmatrix} -3 \\ 5 \end{pmatrix}$

C $\begin{pmatrix} -5 \\ 3 \end{pmatrix}$

D $\begin{pmatrix} 5 \\ -3 \end{pmatrix}$

E $\begin{pmatrix} -5 \\ -3 \end{pmatrix}$

11. There are 24 marbles of assorted colours, red, white and blue, in a bag. One marble is drawn at random from the bag. The probability that it will be red is $\frac{1}{4}$ and that it will be white is $\frac{1}{3}$. The number of blue marbles in the bag is

A   6

B   8

C   10

D   14

E   18

12. Given that $3x - 7 > 10$ and that x is a positive integer, then the solution set of the inequation is

A $\{17, 18, 19, \ldots\}$

B $\{6, 7, 8, \ldots\}$

C $\{5, 6, 7, \ldots\}$

D $\{1, 2, 3, \ldots\}$

E $\{2, 3, 4, \ldots\}$

13. $3x^2 + x - 2 \equiv$

A $(3x - 1)(x - 2)$

B $(3x - 1)(x + 2)$

C $(3x + 1)(x - 2)$

D $(3x - 2)(x + 1)$

E $(3x + 2)(x - 1)$

14. In the cyclic quadrilateral $PQRS$, angle P : angle $R = 5:4$.

Angle $R =$

A 20°

B 80°

C 100°

D 144°

E 160°

15. The area of a circle is $16 \, \text{cm}^2$. The radius, in cm, is

A $\dfrac{4}{\sqrt{\pi}}$

B $\sqrt{\left(\dfrac{8}{\pi}\right)}$

C $\dfrac{2}{\sqrt{\pi}}$

D $\dfrac{8}{\pi}$

E $\sqrt{\left(\dfrac{32}{\pi}\right)}$

16.

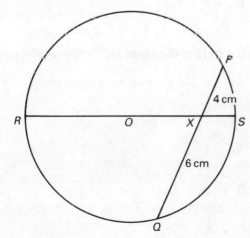

The diameter, RS, of the circle shown is equal to $11 \, \text{cm}$.

Then $XS =$

A $2\frac{2}{11} \, \text{cm}$

B $3 \, \text{cm}$

C $4\frac{2}{5} \, \text{cm}$

D $6\frac{3}{5} \, \text{cm}$

E $8 \, \text{cm}$

17.

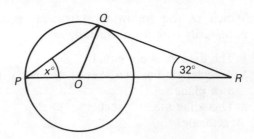

RQ is a tangent to the circle centre O.

Then $x =$

A 16°

B 29°

C 32°

D 34°

E 58°

18.

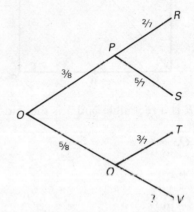

In the given tree diagram the probability is missing from one of the branches. The missing value is

A $\dfrac{4}{8}$

B $\dfrac{5}{7}$

C $\dfrac{4}{7}$

D $\dfrac{3}{8}$

E $\dfrac{2}{9}$

19. Which of the following statements is (are) necessarily true for all rectangles?

1 The diagonals are equal.
2 A rectangle has two axes of symmetry in its own plane.
3 The diagonals bisect the angles of the rectangle.

A 1, 2 and 3

B 1 and 2 only

C 1 and 3 only

D 1 only

E 2 only

20.

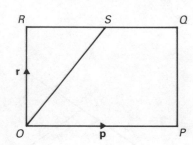

$OPQR$ is a rectangle and S is the mid-point of RQ.
Then $\overrightarrow{OS} =$

A $\mathbf{p} + \mathbf{r}$

B $\mathbf{r} - \frac{1}{2}\mathbf{p}$

C $\frac{1}{2}(\mathbf{p} + \mathbf{r})$

D $\mathbf{p} + \frac{1}{2}\mathbf{r}$

E $\frac{1}{2}\mathbf{p} + \mathbf{r}$

21. When the domain of the function $f: x \mapsto x^2$ is defined as $0 \leqslant x \leqslant 4$, the range of f is

A $-2 \leqslant f(x) \leqslant 2$

B $0 \leqslant f(x) \leqslant 2$

C $-4 \leqslant f(x) \leqslant 4$

D $-16 \leqslant f(x) \leqslant 16$

E $0 \leqslant f(x) \leqslant 16$

22.

x	1	2	3	4	5
Frequency	2	3	1	5	1

The difference between the mean and the mode of the given distribution is

A 4

B 3

C 2

D 1

E 0

23. Given that the matrix $\begin{pmatrix} x & y \\ 5 & 2 \end{pmatrix}$ is singular, then

A $2x - 5y = 0$

B $2x + 5y = 0$

C $xy + 10 = 0$

D $x + y = 10$

E $x - y = 3$

24. Given that the point $(a, 5)$ lies on the curve
$$y = \frac{2}{x} - 3,$$
$a =$

A -4

B $-2\frac{3}{5}$

C $-\frac{1}{3}$

D $\frac{1}{4}$

E 4

25.

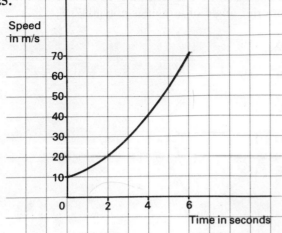

Of the following, the best approximation to the distance covered in the 6 seconds of motion represented by the time–speed graph is

A 140 m

B 200 m

C 210 m

D 240 m

E 420 m

26.

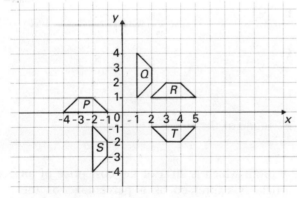

A translation with vector $\begin{pmatrix} -3 \\ -5 \end{pmatrix}$ maps

A region P onto region R

B region R onto region P

C region Q onto region S

D region T onto region P

E region S onto region Q

27. Given that $y = m(x - a)$, then $a =$

A $mx - y$

B $y - mx$

C $\dfrac{mx - y}{m}$

D $\dfrac{y - mx}{m}$

E $\dfrac{y}{m} - x$

28.

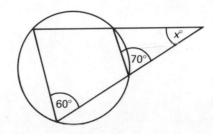

$x =$

A 10

B 30

C 40

D 50

E 60

29. When $x = 4$, $9x^{-\frac{1}{2}} =$

A $\dfrac{1}{6}$

B $\dfrac{9}{16}$

C $\dfrac{9}{4}$

D $\dfrac{9}{2}$

E 18

30.

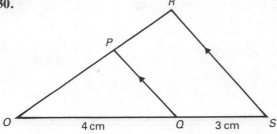

Given that PQ is parallel to RS,

then $\dfrac{\text{area } OPQ}{\text{area } ORS} =$

A $\quad \dfrac{16}{49}$

B $\quad \dfrac{4}{7}$

C $\quad \dfrac{16}{9}$

D $\quad \dfrac{4}{3}$

E $\quad \dfrac{3}{4}$

31. The volumes of two spheres are in the ratio $27:64$. The ratio of their surface areas is

A $\quad 9:16$

B $\quad 3:4$

C $\quad 27:64$

D $\quad 9:8$

E $\quad 27:16$

32. The scale of a map is $1:50\,000$. A length of $2 \cdot 8$ cm on the map represents

A $\quad 14\,000$ km

B $\quad 1400$ km

C $\quad 140$ km

D $\quad 14$ km

E $\quad 1 \cdot 4$ km

33. Which one of the following could *not* be the interior angle of a regular polygon?

A $\quad 155°$

B $\quad 160°$

C $\quad 165°$

D $\quad 170°$

E $\quad 175°$

34.

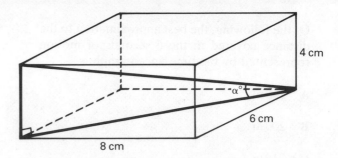

The diagram shows a cuboid with edges 8 cm, 6 cm and 4 cm.
Then $\sin \alpha° =$

A $\quad \dfrac{2}{5}$

B $\quad \dfrac{5}{2}$

C $\quad \dfrac{4}{\sqrt{80}}$

D $\quad \dfrac{2}{7}$

E $\quad$ none of these

35. A profit which is 20% of the cost price of an article is expressed as a percentage of the selling price. This percentage

A $\quad$ is 12%

B $\quad$ is $16\frac{2}{3}$%

C $\quad$ is 20%

D $\quad$ is 60%

E $\quad$ cannot be found

36.

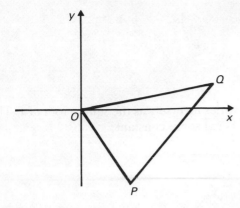

Given that $\overrightarrow{OP} = 2\mathbf{i} - 3\mathbf{j}$, $\overrightarrow{OQ} = 5\mathbf{i} + \mathbf{j}$ then the modulus of $\overrightarrow{PQ}$ is

A $\sqrt{13} - \sqrt{26}$

B $\sqrt{39}$

C $3 \cdot 5$

D 5

E 7

37.

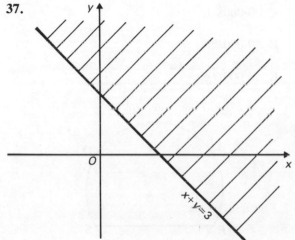

Which of the following is (are) needed to define completely the shaded region of the x–y plane?

1 $x + y > 3$
2 $x > 0$
3 $y > 0$

A   **1** only

B   **1** and **2** only

C   **1** and **3** only

D   **2** and **3** only

E   **1, 2,** and **3**

38.

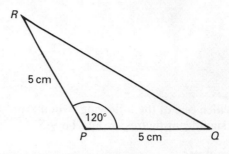

The length, in cm, of QR is

A 10

B $5 \sin 60°$

C $10 \sin 60°$

D $10 \cos 60°$

E $\sqrt{50}$

39.

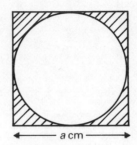

The area, in cm^2, of the shaded region, is

A $a^2 \left(1 - \dfrac{\pi}{4}\right)$

B $a^2 (1 - \pi)$

C $a^2 \left(1 - \dfrac{\pi}{2}\right)$

D $a(a - \pi)$

E $a(a - 2\pi)$

40. A chest of n kg of tea is weighed into p bags each holding r grams. What weight, in grams, of tea is left over?

A $n - pr$

B $n - \dfrac{pr}{1000}$

C $1000n \quad pr$

D $\dfrac{1000n}{pr}$

E $100n - pr$

107

Test 24

Time allowed: 1 hour

1. Which one of the following, when expressed to 3 significant figures, would be 9·55?

   A 0·955

   B 9·505

   C 9·556

   D 9·555

   E 9·546

2. £P is to be divided between Adam and Eve in the ratio $x:y$. The amount, in £'s, which Eve will receive is

   A $\dfrac{yP}{x}$

   B $\dfrac{xP}{y}$

   C $\dfrac{yP}{x+y}$

   D $\dfrac{xP}{x+y}$

   E $\dfrac{P}{x+y}$

3.

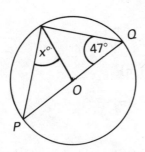

   POQ is a diameter. $x =$

   A 43

   B 45

   C 47

   D 86

   E 94

4. y varies inversely as the square root of x. Then, k being some constant,

   A $y = k\sqrt{x}$

   B $y = \dfrac{k}{\sqrt{x}}$

   C $y = kx^2$

   D $y = \dfrac{k}{x^2}$

   E $\sqrt{y} = \dfrac{k}{x}$

5. $\sqrt{(0\cdot000\,064)} =$

   A 0·08

   B 0·008

   C 0·0008

   D 0·000 08

   E 0·000 008

6.

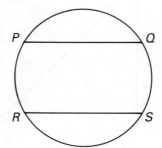

   PQ and RS are equal, parallel chords of a circle of radius 17 cm. The distance between the chords is 16 cm. The length, in cm, of PQ is

   A $2\sqrt{33}$

   B 15

   C $\sqrt{353}$

   D 18

   E 30

7. The solution set of the equation $x^2 + 7x = -6$ is

  A   {0, −7}

  B   {0, 7}

  C   {6, 1}

  D   {−6, −1}

  E   {−3, −2}

8. The angle subtended at the centre of a circle of radius 12 cm by an arc of length 2π cm is

  A   15°

  B   30°

  C   45°

  D   60°

  E   90°

9. The median of the numbers 1, −1, 4, −1, −2, 6, 4, 3, 4 is

  A   −2

  B   2

  C   3

  D   4

  E   6

10. Which *one* of the following matrices is non-singular?

  A $\begin{pmatrix} 6 & 4 \\ -6 & 4 \end{pmatrix}$

  B $\begin{pmatrix} 1 & -2 \\ -1 & 2 \end{pmatrix}$

  C $\begin{pmatrix} 2 & 3 \\ 0 & 0 \end{pmatrix}$

  D $\begin{pmatrix} 1 & 1 \\ 1 & 1 \end{pmatrix}$

  E $\begin{pmatrix} 5 & -10 \\ \frac{1}{4} & -\frac{1}{2} \end{pmatrix}$

11.

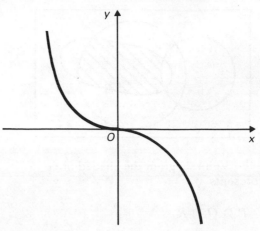

An equation of the curve shown in the sketch could be

  A $y = x^2$

  B $y = -x^2$

  C $y = x^3$

  D $y = -x^3$

  E $y^2 = x$

12. $\dfrac{1}{x + 1} - \dfrac{2}{x + 3} \equiv$

  A $\dfrac{-x + 1}{x^2 + 4x + 3}$

  B $\dfrac{-x + 1}{x^2 + 2x + 3}$

  C $\dfrac{-x + 5}{x^2 + 4x + 3}$

  D $\dfrac{-x - 1}{x^2 + 4x + 3}$

  E $\dfrac{-x - 5}{x^2 + 4x + 3}$

13. $f : x \mapsto 2x^2 + x - 1$, $g : x \mapsto x + 1$. $fg : x \mapsto$

  A $2x^2 + 3x + 2$

  B $2x^2 + x$

  C $(2x^2 + x - 1)(x + 1)$

  D $2x^2 + 5x$

  E $2x^2 + 5x + 2$

14.

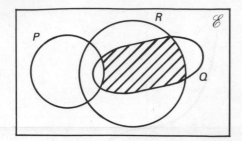

In the Venn diagram the shaded area represents

A $P \cap Q \cap R$

B $(Q \cap R) \cup P$

C $P' \cap Q \cap R$

D $(Q \cup R) \cap P'$

E $(Q \cap R) \cup P'$

15. The probability that, in two throws of a fair die, the total score will be 5 is

A $\dfrac{1}{36}$

B $\dfrac{5}{36}$

C $\dfrac{7}{36}$

D $\dfrac{5}{12}$

E   none of these

16.

Goals scored	0	1	2	3	4	5
Games played	2	5	6	4	2	1

The table gives the goal-scoring record of a football team. The mean number of goals scored per game was

A   2

B   2·1

C   2·2

D   2·5

E   7

17. The transformation whose matrix is $\begin{pmatrix} 0 & 1 \\ 1 & 0 \end{pmatrix}$ is

A   a clockwise rotation through 90°

B   an anticlockwise rotation through 90°

C   a half-turn

D   a reflection in the line $y = x$

E   a reflection in the line $y = -x$

18. $f : x \mapsto$ (the smallest positive integer greater than $\sqrt{x}$).
Then $f[f(35)] =$

A   2

B   3

C   6

D   36

E   37

19.

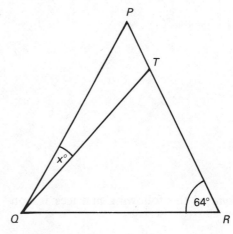

Given that $PQ = PR$ and $QR = QT$, then $x =$

A   12

B   24

C   26

D   32

E   52

20.

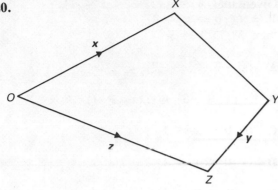

In the figure $\overrightarrow{OX} = \mathbf{x}$, $\overrightarrow{OZ} = \mathbf{z}$, and $\overrightarrow{YZ} = \mathbf{y}$.
Then $\overrightarrow{XY} =$

A $\mathbf{x} + \mathbf{y} + \mathbf{z}$

B $\mathbf{x} - \mathbf{y} - \mathbf{z}$

C $\mathbf{x} - \mathbf{y} + \mathbf{z}$

D $-\mathbf{x} - \mathbf{y} + \mathbf{z}$

E $-\mathbf{x} + \mathbf{y} - \mathbf{z}$

21. X and Y are sets and $n(X) = 80$, $n(Y) = 10$.
Then the least and greatest values of $n(X \cup Y)$
are, respectively

A 8 and 800

B 70 and 80

C 10 and 90

D 80 and 90

E 70 and 90

22. Which of the following is/are true for all
rhombuses?

1 The diagonals are equal.
2 The diagonals are perpendicular.
3 The diagonals bisect the angles.

A All of them

B 2 only

C 3 only

D 1 and 2 only

E 2 and 3 only

23.

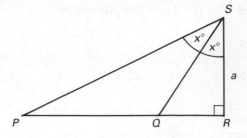

The length, in cm, of PQ is

A $\dfrac{a \tan x°}{\cos x°}$

B $a \tan 2x°$

C $a \tan 2x° - a \tan x°$

D $a \sin 2x° - a \sin x°$

E $a \cos 2x° - a \cos x°$

24.

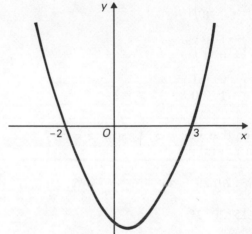

The curve shown could be part of the graph of

A $y = x^2 - x - 6$

B $y = 6 + x - x^2$

C $y = x^2 + x - 6$

D $y = 6 - x - x^2$

E $y = x^2 - 6$

25. Given that $(2^{x-1})^2 = 512$, then $x =$

  A   4

  B $5\frac{1}{2}$

  C $\pm 5\frac{1}{2}$

  D -2

  E   4 or -2

26. Which *one* of the following matrices represents a rotation through 90° in the anti-clockwise sense?

  A $\begin{pmatrix} 0 & 1 \\ -1 & 0 \end{pmatrix}$

  B $\begin{pmatrix} 1 & 0 \\ 0 & -1 \end{pmatrix}$

  C $\begin{pmatrix} 0 & -1 \\ 1 & 0 \end{pmatrix}$

  D $\begin{pmatrix} -1 & 0 \\ 0 & 1 \end{pmatrix}$

  E $\begin{pmatrix} 0 & -1 \\ -1 & 0 \end{pmatrix}$

27. Given that $x:y = 3:4$ and $y:z = 5:7$, then $x:y:z =$

  A $3:20:28$

  B $15:20:21$

  C $15:20:7$

  D $21:28:20$

  E $15:20:28$

28. The heights of two circular cylinders are in the ratio $5:2$ and their radii are in the ratio $2:3$. Then the ratio of their volumes is

  A $25:9$

  B $5:3$

  C $15:4$

  D $10:9$

  E $25:6$

29. Given that $X = \{x : -3 < x < 1\}$, $Y = \{x : 0 < x < 2\}$, then $X' \cap Y =$

  A $\{x : -3 < x < 2\}$

  B $\{x : x \leqslant -3\} \cup \{x : x \geqslant 1\}$

  C $\{x : 0 < x \leqslant 1\}$

  D $\{x : 0 < x < 1\}$

  E $\{x : 1 \leqslant x < 2\}$

30. The greatest number of cubes of edge 50 cm that can be packed into a cubical box of internal edge 2 m is

  A   4

  B   8

  C   16

  D   64

  E   256

31.

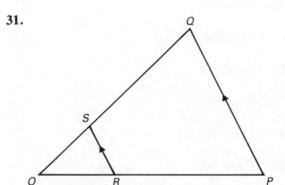

In the figure $OR:RP = 1:2$ and RS is parallel to PQ. Given that $\overrightarrow{OP} = \mathbf{p}$ and $\overrightarrow{OQ} = \mathbf{q}$, $\overrightarrow{RS} =$

  A $\frac{1}{2}(\mathbf{q} - \mathbf{p})$

  B $\frac{1}{2}(\mathbf{p} - \mathbf{q})$

  C $\frac{1}{3}(\mathbf{q} - \mathbf{p})$

  D $\frac{1}{3}(\mathbf{p} - \mathbf{q})$

  E $\frac{1}{3}(\mathbf{p} + \mathbf{q})$

32. Given that $5x + 9 > 24$ and $7x + 6 < 27$, then a possible value of x is

A   4

B   3

C   2

D   1

E   none of these; there is no value for x

33.

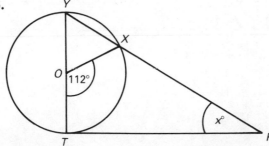

PT is the tangent at T to the circle, centre O, and YOT is a diameter.

$x =$

A   34

B   39

C   44

D   56

E   68

34. The scale of a map is 1 cm to 1½ km. A rectangle on the map has sides 3 cm and 6 cm. The actual area, in km², of this rectangle on the ground is

A $40\frac{1}{2}$

B   27

C $22\frac{1}{2}$

D   12

E   8

35. A card is to be drawn from an ordinary pack of 52 playing cards and, at the same time, a die is to be thrown. The probability that a heart will be drawn and a six will *not* be thrown is

A $\dfrac{1}{24}$

B $\dfrac{1}{8}$

C $\dfrac{5}{24}$

D $\dfrac{5}{8}$

E $\dfrac{19}{24}$

36.

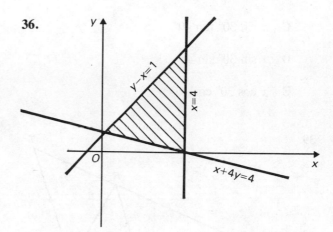

The coordinates of any point lying within the shaded region must satisfy the following inequalities

A $y - x < 1,\ x < 4,\ x + 4y < 4$

B $y - x < 1,\ x < 4,\ x + 4y > 4$

C $y - x > 1,\ x < 4,\ x + 4y < 4$

D. $y - x > 1,\ x < 4,\ x + 4y > 4$

E $y - x < 1,\ x > 4,\ x + 4y < 4$

37. The modulus of the vector $\begin{pmatrix} -3 \\ 5 \end{pmatrix}$ is

A   4

B -4

C $\sqrt{34}$

D $\pm\sqrt{34}$

E   2

113

38.

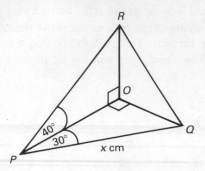

OPQ is a horizontal plane with $\angle POQ = 90°$ and OR is vertical.
$OR =$

A $\quad x \cos 30° \sin 40°$

B $\quad x \cos 30° \tan 40°$

C $\quad x \sin 30° \tan 40°$

D $\quad x \sin 30° \sin 40°$

E $\quad x \cos 30° \cos 40°$

39.

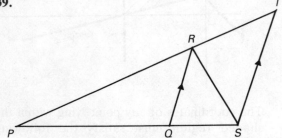

Given that QR is parallel to ST, which of the following statements is/are true?

1 Triangles PQR and PST are similar
2 Triangles RQS and RST are similar
3 Triangles PQR and PRS are similar

A $\quad$ **1** only

B $\quad$ **1** and **2** only

C $\quad$ **1** and **3** only

D $\quad$ **1, 2** and **3**

E $\quad$ none of them

40. Given that $-5 \leqslant x \leqslant 2$ and $y = -x$, which one of the following must necessarily be true?

A $\quad y < -5$ or $y \geqslant 2$

B $\quad y \leqslant -5$ or $y > 2$

C $\quad -5 < y \leqslant 2$

D $\quad -2 \leqslant y \leqslant 5$

E $\quad -2 \leqslant y < 5$